Land drainage

Longman Handbooks in Agriculture

Series editors:

C. T. Whittemore
K. Simpson

Books published

C. T. Whittemore: *Lactation of the dairy cow*
C. T. Whittemore: *Pig production – the scientific and practical principles*
A. W. Speedy: *Sheep production – science into practice*
R. H. F. Hunter: *Reproduction of farm animals*
K. Simpson: *Soil*
J. D. Leaver: *Milk production – science and practice*
J. M. Wilkinson: *Beef production from silage and other conserved forages*

Land drainage

E. Farr
and W. C. Henderson

Longman
London and New York

Longman Group Limited
Longman House, Burnt Mill, Harlow
Essex CM20 2JE, England
Associated companies throughout the world

Published in the United States of America
by Longman Inc., New York
© Longman Group Limited 1986

First published 1986

British Library Cataloguing in Publication Data
Farr, E.
    Land drainage. – (Longman handbooks in
    agriculture)
    1. Drainage
    I. Title   II. Henderson, W. C.
    631.6'2   TC970

    ISBN 0-582-45007-1

Library of Congress Cataloging in Publication Data
Farr, E., 1938 –
    Land drainage.

    (Longman handbooks in agriculture)
    Includes index.
    1. Drainage.   I. Henderson, W. C., 1932–
II. Title.   III. Series.
TC970.F34   1986   627'.54   85–9736
ISBN 0-582-45007-1 (pbk.)

Produced by Longman Group (FE) Limited
Printed in Hong Kong

Contents

Preface xi

Part one   **The physical basis of land drainage   1**

Chapter 1   **The hydrologic cycle   2**
Precipitation   2
Water circulation   3
Pathways for rainwater   3
The effects on agriculture   8
Soil drainage problems   9
Classifying soil drainage problems   10
Land drainage requirements   11

Chapter 2   **Soil water and plant growth   12**
Soil conditions for plant roots   12
The functions of absorbed water   13
The transpiration ratio   14
The amount of soil water transpired   14
Soil water   15
Soil air   15
Soil voids   16
Soil water tension   16
The effects of pore size on soil fertility   17
Soil drainage status   18
The plant environment   19
Land use   19
The effects of poor soil drainage on agriculture   20
The general effect of restricted soil drainage   22
Grassland plants   22
Soils   23
The need for land drainage   23
Assessing the soil drainage status   24
The need for conservation of wildlife   25

Chapter 3   **Geological processes affecting land drainage** **26**
Some definitions   26
The characteristics of rocks   27
Rock weathering   29
Rock waste   30
Water-transported drifts   31
Ice-transported drifts   31
Wind-transported drifts   34
Soil-forming processes   35
Clay minerals   36
The effects of geological processes   37
The use of maps   37

Chapter 4   **Mineral soils** **39**
Soil profiles   39
The characteristics of soil profiles   40
Soil horizons   40
Soil texture   41
Assessing soil texture   41
Soil structure   45
Soil pore spaces   50
Soil induration   51
Soil compaction   52
Soil permeability   52
Soil colour   52
Iron ochre in soils   53
Soil organic matter   54
Stones in soil profiles   54
Soil classification   54
Wetland soils   56
Examining a soil profile   58
Assessing profile hydraulic conductivity   60
Measuring soil properties   61
Drainage problems associated with profile characteristics   62
Soil management   63

Chapter 5   **Organic soils** **64**
Organic matter in soils   64
Classifying organic soils   64
Peat accumulation   65
The influence of climate   65
The influence of topography   67

Raised mires 68
Raised mire profiles 69
The present environment 69
Coastal peatlands 70
The influence of springs 70
The location of peatlands 70
Soil properties 71
The composition of peat 72
Humification 73
Soil texture 74
Soil structure 75
Peat shrinkage 75
Conserving organic soils 76
Profile permeability 77
The value of peat drainage 77
The special difficulties of peatland drainage 78

Chapter 6   **Topography and land drainage 79**
Factors which shape the landscape 79
The importance of time 83
Basic landform shapes 83
The relationship between topography and soil drainage
problems 85
High groundwater tables 88

Chapter 7   **Rainfall 89**
Air masses 89
Rainfall 89
Climatic zones 90
Soil drainage requirements 92
The critical season 93
Rainfall amounts 93
Choice of return frequency 94
Rainfall intensity 94
The area affected by rainfall events 95
The design rainfall rate 95

Chapter 8   **Surface water 98**
Channel discharge 98
The relationship between rainfall and channel discharge 101
Channel erosion 102
Base levels 103

River meanders and flood plains   103
River flood plains   105
Drainage design factors   106
Drainage in river flood plains   106

Chapter 9   **Groundwater   108**
Rock porosity   108
Rock permeability   109
Aquifers and aquicludes   109
The properties of aquifers   109
Surface aquicludes   114
Groundwater pressure   115
The effect of rock formations below an aquifer   116
Dipping aquifers as a source of water   116
The properties of springs   118
Artesian springs   120
Identifying spring seepages   125
Drainage works to control groundwater   126

Part two   **Practical land drainage   127**

Chapter 10   **Ditches   128**
Advantages and disadvantages   128
Functions   128
New ditches   130
Ditch design   131
Design factors for ditches   131
Ditch excavation   137
Bank stabilisation   138
Culverts   139
Ditch improvement   140
Ditch maintenance   143
Consultations   144

Chapter 11   **Underdrainage   145**
Advantages and disadvantages   145
Types of underdrainage   145
Design requirements   147
Drainflow   147
Design drainage rate   147
Choosing a design drainage rate   148
Calculating design drainage rates   153

Pipe discharge capacity   153
Choice of pipe size   156
Layout, depth and spacing of field drains   160
Workmanship and materials   161

Chapter 12   **Subsoiling, mole drainage and permeable fill   175**
Perched water tables   175
Soil treatment   175
Mole drainage   176
Design factors for mole drainage   179
Subsoiling   181
Permeable fill   187

Chapter 13   **Low cost drainage   190**
Difficult sites for land drainage   190
Low-value land   190
Hill drains   191
The marginal areas   193
Difficult drainage in more fertile areas   196
Soils that are difficult to drain   196
Sites that are expensive to drain comprehensively   197

Chapter 14   **Pumped drainage   198**
Low-lying sites   198
The size of the project   198
Project feasibility   199
Physical limitations   199
Assessing project feasibility   203
Preliminary site improvements   203
Siting the pump   204
The pump workload   205
The types of water pumps available   205
Selecting the most suitable pump   206
Controlling a pump   207
Protecting the pump   207
The choice of drainage system   208
Open channel systems   208
Direct pumping of underdrainage systems   209
Layout for underdrainage   210
Economic factors   210

Chapter 15   **Drainage design   211**
Drainage layout   211

Layout for ditches 211
Layout for underdrainage 212
Choice of layout 215
Drain spacings 217
Drain depth 217
Classifying drainage problems 217
Classifying profile permeability 218
Controlling a high groundwater table in a surface aquifer 219
Controlling a perched water table in a surface aquifer 224
Controlling a perched water table in a surface aquiclude 225
Controlling overland surface flow 225
Controlling spring seepage 225
Miscellaneous problems presenting special difficulty 231

Part three **A systematic approach to field drainage 237**

Preliminary survey 238
Detailed survey 238
Maintenance work 240
Drainage design 240
Drainage work 241

Glossary 242

Index 249

Preface

Land drainage is an important aspect of farming in many countries of the world. Its influence on plant growth, soil conditions and management operations is widely appreciated and examples of practical importance can be traced back several hundred years. The subject is complex, involving a variety of different disciplines which include Soil Science, Meteorology, Geology, Hydrology, Engineering and Agriculture. This list is by no means exhaustive and the practical considerations in the field arising from the great variability in soil conditions and other factors often defy simple description or analysis.

Thus, the precise nature and effect of a particular drainage problem depends on a whole variety of circumstances and conditions so that though many of the answers may be available, they are rarely contained within one text but rather are dealt with in a series of books, the particular emphasis of each reflecting the authors' interests or experience.

The aim of this book is to provide, in practical and theoretical terms, a broad view of the subject of land drainage as a whole. It has as its origins the difficulties experienced by the authors in their early attempts to solve drainage problems and in seeking guidance from the available literature. It is their intention, not only to find a more acceptable balance between sound drainage techniques and design theories but also to provide some basic knowledge of the natural processes that influence land and soil fertility. It sets out the basic scientific principles of related subjects that may stimulate the student to widen his interest in land drainage matters. It describes simply, and with a minimum of calculation, the principles of scientific designs and where to apply them. Perhaps most important of all, it includes a comprehensive guide to practical land drainage techniques since it is aimed for the field worker as well as the student, to the farmer and administrator, as much as the consultant or land owner.

Although they are grateful to their employers, the Department of Agriculture and Fisheries for Scotland and The North of Scotland College of Agriculture, the authors stress that they accept sole responsibility for the views and information contained in the book. They have drawn on their practical field experience and a great body of published material from a wide variety of sources. Much of the data has been re-constructed and interpleted but where information has been taken directly from a single source, it has been acknowledged.

They are indebted to a large number of friends and colleagues for support, encouragement and discussion but, in particular, wish to record their thanks to Mr T G Brownlie, Mr A Edwards, Mr C Mackay and Mr G M B Redpath (D.A.F.S.), to Mr I Brown, Mrs L McKen and Dr J H Topps (N.O.S.C.A.) and to Mr B M Shipley (Macaulay Institute for Soil Research). They are also grateful to the publishers and especially to the Series Editor, Mr Ken Simpson. Whilst accepting full responsibility for any errors they are much indebted to Mrs E Watt (N.O.S.C.A.) for typing many drafts and modifications.

Part one **The physical basis of land drainage**

Chapter 1 **The hydrologic cycle**

The circulation of water through the atmosphere and about the earth's surface affects the soil in a number of ways. In some circumstances accumulations of excess water will occur, leading to soil drainage problems.

Precipitation

With the exception of small amounts derived from volcanic steam or from sea spray, all water found in soil is derived, directly or indirectly, from the atmosphere. The atmosphere carries water vapour because of the effect of **evaporation**. Wherever air comes in contact with liquid water there is a tendency for water molecules to pass from the body of liquid and mix with the atmosphere. This process can continue until the air near the water body is saturated with water vapour and the two media are then in equilibrium. Warm air can absorb more water molecules than cold air. Atmospheric circulation carries the air with its absorbed water molecules about the globe, often in a manner which causes the air to cool. Any cooling of the air mass reduces its capacity to hold water vapour and if the saturation point is reached any further cooling forces the excess water vapour to **condense** as droplets of liquid water which become visible as cloud or mist. When the droplets coalesce into larger drops they fall to the ground as **precipitation** – a term used to include all rain, hail, sleet or snow that lands on the surface of the earth. For the purpose of discussing land drainage the term **rainfall** is used but this includes the other forms of precipitation. Rainfall is measured in millimetres per unit time. A statement like '10 mm of rain fell in 24 hours' means that, had the rainwater remained on the surface where it fell, the whole surface would have a cover of water 10 mm deep. This represents a volume of 100 m^3 of water on a hectare of land.

Water circulation

The fate of water falling on the surface is varied. Some may return to the atmosphere directly, the rest may remain on the surface or flow downwards into the soil to become **soil water**. The term **groundwater** is also appropriate, but this includes water held in rocks below the soil. Plant activity can return some soil water to the atmosphere and in some situations groundwater may be returned to the surface by various pathways through the soil or rocks. Water on the surface joins the surface drainage system (the streams and rivers) and most of it will ultimately reach the sea. This continuous circulation of water from ocean to atmosphere, to land surface, to ocean, is called the **hydrologic cycle**. Consideration of the hydrologic cycle in some detail identifies the various causes of soil drainage problems. A diagrammatic illustration of water circulation is shown in Fig. 1.1.

Pathways for rainwater
Evaporation

In most conditions some of the incident rainfall will evaporate from the ground surface or from the aerial parts of plants and return directly to the atmosphere, a process which may be observed when hot sunshine follows a shower. More often it is noticed only as a drying effect which can occur at any time of year when the air is not saturated. The evaporation rate is controlled by the temperature and **relatively humidity** of the air. Air totally saturated with water vapour has a relative humidity of 100 per cent. If the relative humidity is less than this the atmosphere can absorb further water and has a drying effect on the soil. Thus the proportion of rainfall returned directly to the atmosphere is greatest in a hot, dry climate and least where conditions are cool and wet.

Transpiration

Growing plants remove much water from the soil. Some of the water absorbed by the plant is incorporated into tissues but most of it passes through the pores on aerial parts of the plant and enters the atmosphere as water vapour. This vital process is called **transpiration**. The rate of transpiration depends upon the number of plants in a given area, their size and their rate of growth. Transpiration removes little or no water from the soil during the dormant season but rapidly builds up to a peak in the growing season when plants are most active. The highest rates occur when the leaf canopy is complete and a vigorous young forest plantation can remove as much as 1 000 mm of rainfall in a year if

Figure 1.1   The hydrologic cycle

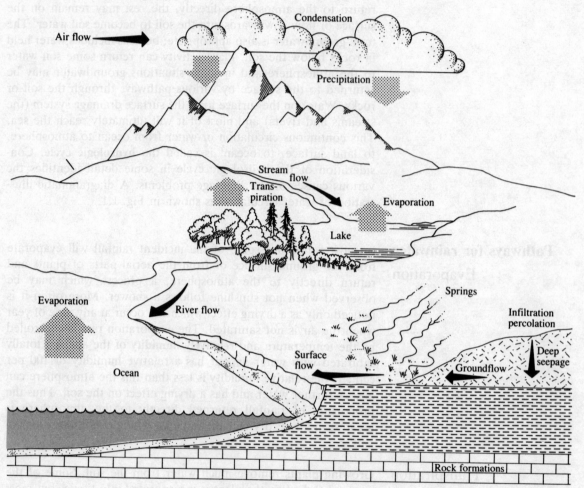

other factors allow. Combined evaporation and transpiration remove considerable amounts of water from the soil. For practical purposes the two effects are identical and are measured together as a total water loss due to **evapotranspiration**.

Flow into the soil

Nearly all land surfaces have a cover of **soil** which is the result of physical, chemical and biological forces acting on the surface rocks over a long period. A vertical cross-section through the soil as seen on the side of a newly dug pit is called a **soil profile**. The profile consists of a number of more or less distinct horizontal layers or **horizons**. From the surface downwards the most prominent horizons are the **topsoil**, usually darkest in colour, the **subsoil** and, lightest in colour, the **soil parent material**. The soil horizons rest on solid or decomposed rock. Most of the rainfall flows into the soil through the spaces between soil particles. The amount of water that enters the soil in mm/day is called the **infiltration rate** and its value is determined by the rainfall rate, the nature and condition of the surface soil layers, the slope of the ground and the amount of vegetation cover. The topsoil where most plant roots are to be found usually has different properties, notably by allowing freer passage to soil water, from the subsoil layers beneath and consequently soil water may be delayed as it moves into the subsoil. The rate at which soil water moves from the topsoil to the subsoil is called the **percolation rate**. Methods are available for measuring the infiltration rate but the percolation rate can be deduced only from observations of soil water movements.

Groundwater

Rainwater entering a dry soil adheres to soil particles as a film of water or is absorbed by spongy organic substances until all the smaller soil spaces are filled. When this requirement is met in the topmost layers further additions of water are free to pass downwards through the larger soil spaces to lower layers so that each soil horizon is replenished with soil water in turn from the surface downwards. Rainfall in excess of the amount that can be retained in soil horizons then passes on to deeper zones and into voids in the rock formations below the soil. This movement to depth beyond the reach of most plant roots is called **deep seepage** and it continues until the water reaches a level below which all voids are already filled with water. The upper surface of this completely saturated zone is called the **groundwater table**. Where there are no horizontal escape pathways continuing rainfall fills more and more of the spaces causing the groundwater table to rise progressively through the subsoil and then the topsoil until it reaches the surface. The soil is then completely saturated with water.

Surface water

When areas of flat land become saturated to the surface, further rainfall remains on the surface as **ponded water** which at first occupies only minor hollows but larger bodies of water develop if the rainfall continues. Surface ponding may also occur when the soil is not saturated if, for a time, the rainfall intensity (precipitation rate) exceeds the rate at which water can enter the soil (infiltration rate). This type of ponding, however, can disappear soon after the rain storm eases unless the soil surface is particularly impervious as it can be, for example, after heavy field traffic. On a sloping surface or where the accumulation of ponded water is considerable, a **surface flow** develops either as **sheet flood**, which is rare in maritime countries, or as a series of **runnels** or **rivulets** which run into the surface drainage pattern of streams and rivers. In a similar way when soil waterlogging has reached the base of the topsoil layers or when the infiltration rate exceeds the percolation rate soil water accumulating within the topsoil layers may move laterally through the topsoil as an **interflow**. This depends on some degree of porosity in the topsoil and where the topsoil is impermeable then interflow cannot occur. At some point downslope the interflow will come to the surface either as ponding or surface flow. Any water flowing across land at or near the surface can cause soil erosion and gullying, especially when the soil is loose and the flow is rapid.

Springs

Rock formations below the subsoil are very variable and range from types that are totally impervious to others that are so porous that they offer little resistance to water movements. In the latter case the groundwater table may not rise with additions of rainfall because the excess water will flow elsewhere through the rocks. This is particularly important on elevated sites. Any general horizontal or upwards movement of groundwater through the rocks or the subsoil layers is described as **groundflow** and its usual effect is to return groundwater to the surface usually at some lower part of the landscape where it adds to the surface flow. Sites where groundwater returns to the surface remain wet in dry weather and are variously described as **springs**, **issues**, or **seepages**.

Surface slopes

The combined effects of surface flow, interflow and groundflow shed rainfall off the hills and slopes and cause water to be concentrated in the valleys and hollows. During each period of rainfall elevated soils must accept less, and soils in collecting sites more, than the incident rainfall. Furthermore, groundflow often causes continued accumulation of water in collecting sites after the rainfall has stopped. This has a significant effect on the groundwater table which can be many metres below the surface of elevated parts of the landscape and at or near the surface in lower parts. Where there are natural basins or river channels the groundwater table stands above the surface as a permanent feature. Thus the general level of the groundwater table is determined by the **topography** or general shape of the land surfaces present. This is shown in Fig. 1.2.

Figure 1.2   *The effect of topography on the groundwater table*

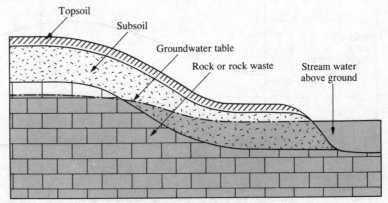

Topsoil Subsoil Groundwater table Rock or rock waste Stream water above ground

Movement of the groundwater table

In dry weather water is lost from the soil by evapotranspiration and perhaps also by groundflow so that the zone of saturation descends and the soil tends to dry out. In maritime climates rainfall is more or less evenly distributed throughout the year but evapotranspiration is predominantly a growing-season process. This causes an annual rising and falling of the groundwater table about a median level in the soil or underlying rocks. Superimposed on this pattern are shorter term fluctuations due to individual periods of rainfall as shown in Fig. 1.3.

The return flow

In a maritime climate annual gains from rainfall exceed losses by evapotranspiration and the difference between their values is described as **excess rainfall**. This is shed from the land by means of lateral (horizontal) flow and becomes the streams and rivers of the natural surface drainage. Sometimes the flow enters a lake but these have permanent overflows which also add to river flow. There will be some evaporation losses from the surface of lakes and rivers but all the remaining flow is ultimately returned to the

Figure 1.3   The effect of climate on the groundwater table

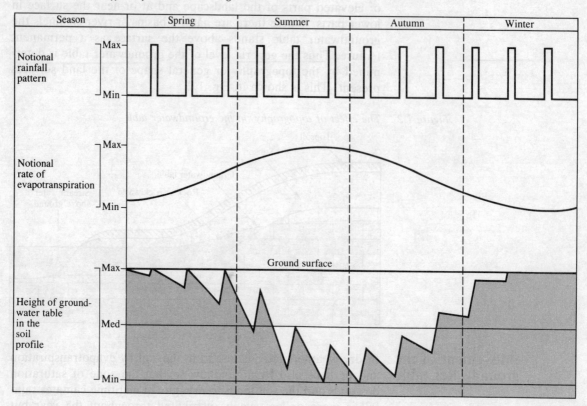

sea to complete the hydrologic cycle. In a dry climate the evapo-transpiration equals rainfall each year so that there is no excess rainfall and no permanent surface drainage.

The effects on agriculture

The influence of the local climate on the hydrologic cycle has far-reaching consequences for the food crops that can be grown. In arid lands crops can be grown only if water is added to the soil by a system of **irrigation** to augment natural rainfall. In a region with excess rainfall it may be necessary to increase the natural loss of water from the soil by **land drainage** works. Because of variations of soil type, rock type, surface shape and climate, several types of drainage problems are found.

Soil drainage problems
On level sites

Soil water pathways for level sites are shown in Fig. 1.4.

On a level site receiving only incident rainfall and where the soil **is** highly permeable:

1 When evapotranspiration exceeds rainfall the soil will tend to dry out and the groundwater table will fall.
2 When rainfall exceeds evapotranspiration the soil will get wetter and the groundwater table will rise.

On a level site receiving only incident rainfall and where the soil **is not** highly permeable:

3 When rainfall exceeds infiltration, water will tend to collect on the surface as long as the rates are maintained.
4 When infiltration exceeds percolation, water will tend to collect in the topsoil as long as the rates are maintained.
5 When water is delayed near the surface the effect of rainfall on groundwater table movements is reduced.

Figure 1.4   Soil water movements – surface level

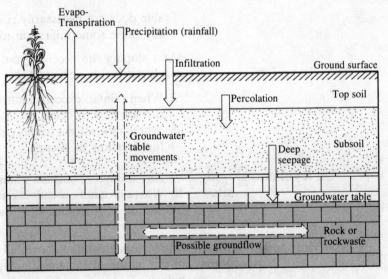

On sloping sites

Soil water pathways for sloping sites are shown in Fig. 1.5.

On a sloping site receiving only incident rainfall and where the soil **is** highly permeable:

6 When rainfall exceeds evapotranspiration the groundwater

Figure 1.5 *Soil water movements – surface sloping*

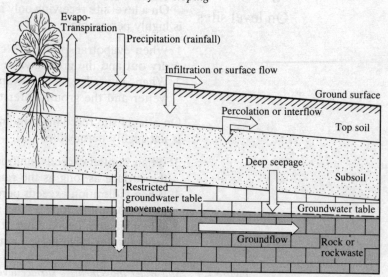

table does not necessarily rise in response because groundflow may cause some migration to another site.

On a sloping site receiving only incident rainfall and where the soil **is not** highly permeable:

7 When rainfall exceeds infiltration there will be a tendency for surface flow to occur.
8 When infiltration exceeds percolation there will be a tendency for interflow to occur.

On level and sloping sites:

9 Any inflows of surface water may make the soil wet.
10 Where groundflow causes groundwater to return to the surface the soil may remain wet in any kind of weather.

Classifying soil drainage problems

The factors identified above fall into four natural groups in terms of the nature of the soil drainage problem. These are described below:

1 The interaction of climate and soil characteristics Factors (1) to (5) above are all concerned with the direct effects of rainfall and evapotranspiration. Rainfall varies widely with time and over the surface, evapotranspiration depends on air temperature and humidity as well as the cover of vegetation while soils have a wide range of values of permeability. In general, the greater the rainfall, the lower the temperature, the poorer the soil permeability, the greater is the drainage problem. Arid deserts and blanket peat mires are the extremes of such factors. Accumulation of rainfall on the surface or in near surface soil horizons due to factors (3), (4), (7) and (8) produces a zone of saturation overlying horizons which are not saturated. This is described as a **surface water problem** or a **perched water table.** This is quite different from a groundwater table which implies total saturation below that level. For this reason the term 'water table' should not be used without a prefix.

2 The effect of surface topography Surface flow, interflow and ground flow due to factors (6), (7) and (8) are dictated by the shape of the land surface which causes water to collect in natural hollows and has the effect of bringing the groundwater table nearer to the surface. Where crop growth is inhibited it is caused by a **high groundwater table** and in this situation soils with the same rainfall will have drainage characteristics that are influenced by their position on the landscape.

3 Surface inflows Some sites are affected by surface inflows from nearby steep land and occur as a result of factors (6), (7), (8) and (9) in situations where the source area cannot be improved.

4 The effect of groundflow The combined effects of factors (6) and (10) are the least predictable in drainage practice. Ground-flow may cause springs to occur on any type of site independently of climate, topography or soil properties.

Land drainage requirements Land drainage requires the recognition and alleviation of the four problems of excess soil water detailed above which may occur singly or in any combination. The basic requirements of drainage improvements may involve:

- Improvement of the permeability of soil horizons.
- Control of the groundwater table in the soil profile.
- Interception of surface inflows.
- Interception of groundflow so that it cannot reach the surface.

Chapter 2 Soil water and plant growth

All plants need a supply of water to carry out their vital functions. For land plants this is obtained from the soil and the amount of water available depends on the local climate and soil properties. With a wide range of such factors, wild plants have become adapted to survive in a particular environment and do less well in a different environment. Crop plants are equally restricted. Where crops are to be grown in dry soils it is usually necessary to irrigate. At the same time, however, soils that are too wet can inhibit crop growth. In addition to the direct effect wet soil has on plants it is also susceptible to damage by field traffic which further reduces fertility.

Soil conditions for plant roots

Most plants require a soil that is sufficiently friable to permit growing root systems to ramify throughout the whole mass to obtain water and nutrients. These are extracted by means of the finest branches, the **root hairs**. Root hairs are delicate, single cells with thin walls which come in close contact with soil particles (see Fig. 2.1). Osmotic forces cause the soil water to pass through this semi-permeable membrane towards the more concentrated solution of the cell sap. To carry out their water extraction function the cells of root tissue respire and require a supply of oxygen. This is obtained directly from air occupying voids in the soil mass. Hard, compacted soils physically inhibit root growth while both compacted and saturated soils prevent root growth because there is no air supply. Where soils are exceptionally loose and have a large proportion of air spaces the plants may suffer from lack of moisture in a dry season. Most crop plants are affected to some extent if soil horizons in the rooting zone become saturated and remain so from 2 to 20 or more days depending on species and stage of growth. The natural plants of wet soil survive because air passes to root cells from the atmosphere through air passages within the plant tissue (e.g. rushes) or because they have an extensive shallow root system growing above the zone of saturation.

The functions of absorbed water
Transpiration

Transpiration causes absorbed water to pass from root tissue to all parts of the plant and ultimately the greater part of it passes out through the stomata on the leaf surface and is lost as water vapour. This loss of water causes a suction force in the plant vascular system which ensures a constant supply of water to cool the tissues and to reduce the risk of wilting.

Turgidity

Water is needed in quantity to provide **turgidity**. Many of the aerial parts of plants do not have specialised supporting tissue and the required fabric strength is obtained by osmotic pressure inside each cell which is maintained by water drawn from the vascular transpiration stream. The whole structure can then withstand the stresses of wind and rain but any deficiency of soil water can result in loss of turgidity and the plant wilts.

Transport

Plants need water as a medium of **transport** to move materials from one part to another. All higher plants have vascular systems through which materials are moved in aqueous solution. Carbohydrates manufactured in the leaves are transported to where they are needed, usually in the growing points, or to where they are stored, as for example in potato tubers. Essential plant nutrients like nitrates, phosphates and potassium are absorbed and transported as ions in soil water.

Photosynthesis

Water is needed as an ingredient for photosynthesis, the process by which green plants convert the energy of sunlight into a form that can be used by living tissues. By a complex process within green cells, carbon dioxide from the atmosphere is combined with water from the soil to form a carbohydrate, like glucose, with the release of gaseous oxygen. It is likely that all atmospheric oxygen is the result of plant activity.

The transpiration ratio

The quantity of water required to carry out such functions is very large and, although each process must be supplied if growth is to be satisfactory, most of the water is needed for transpiration. The amount of water needed to produce a given amount of plant tissue is called the **transpiration ratio**:

$$\text{Transpiration ratio} = \frac{\text{Amount of water transpired}}{\text{Amount of dry plant material produced}}$$

The amount of soil water transpired

Plants can transpire as much as 6 mm of rainfall equivalent per day in the growing season, a figure which represents 60 000 litres/ha. The amount of rainfall needed to meet the plant water requirements depends on the length of the growing season, the temperature and humidity of the air, the exposure and wind strength. Since transpiration cannot easily be separated from evaporation they must be considered together as evapotranspiration. It is useful also to compare **actual evapotranspiration** – the losses of soil water that measurement would record – and **potential evapotranspiration**, the amount of water that could pass into the atmosphere if the amount of soil water were not a limiting factor. In the hot, humid equatorial rainforests the potential evapotranspiration of about 1 500 mm/year is actually achieved by rainfall of 2 000 mm or more with the excess rainfall becoming river flow. Even higher values of potential evapotranspiration occur in the dry tropical zones and this may occur in well-watered valleys, but the usually small rainfall values are a severe limiting factor. In higher latitudes the potential evapotranspiration declines with lower temperatures and plant requirements are usually met in the maritime regions as in France or the eastern part of the United States where a potential value of about 550 mm is offset by total rainfalls of about 750 mm. In the same latitudes but towards the continental interiors, rainfall declines to about 200 mm, well below the potential water loss. Further north, in Canada or Scandinavia, the potential evapotranspiration drops to about 150 mm so that the soils are rarely deficient of water. Crop plants exhibit different water demands. Crops like cereals transpire about 500 mm of soil water in a season while actively growing forests reach a value of about 1 000 mm if other factors permit.

Soil water The amount of water in a soil can be calculated as the **soil water content**:

$$\text{Water content} = \frac{\text{weight of water in soil}}{\text{weight of oven dry soil}} \times \frac{100}{1}$$

The amount of water in a soil profile varies with the seasons as the balance of inputs and outputs change. It also varies from soil to soil depending on the proportion of voids within the soil mass. Any soil profile will be saturated to the surface immediately after a major period of heavy rainfall. When rainfall ceases, provided the soil can drain freely, any **gravitational water** will drain from the larger soil voids in a few days or less and when all gravitational water has been lost the soil is said to be at **field capacity**. When evapotranspiration is the dominant factor the soil dries out, the soil water content falls below field capacity and a **soil water deficit** occurs. As the water loss continues the soil water deficit increases to a point where plants can no longer extract sufficient water to keep cells turgid so that their tissues collapse. This is called the **wilting point**.

Soil air Every living plant cell needs a continuous supply of gaseous oxygen for respiration and the resulting carbon dioxide must be able to escape. This exchange of gases presents little difficulty in the aerial parts of plants, even in the largest structures, because a network of air spaces connects through lenticels to the atmosphere. In species like rice and rush the network of air spaces extends downwards into the root system, but for the majority of plants the essential gaseous exchange to maintain an oxygen supply to root tissue must occur in air-filled soil voids and these voids must form a continuous network connecting to the atmosphere. For this reason a growing root system must obtain from the soil, at the same time, oxygen, water and dissolved plant nutrients. This is possible only where there is a range of soil void sizes, some containing air and the others filled with water.

Soil voids Soil voids or **pore spaces** may be classified according to their diameters. The influence they have on soil properties and on the behaviour of soil water is determined by the distribution of the various sizes of pores throughout the soil profile and this is a fundamental concept for crop production. It is useful to recognise the following size groupings:

- **macropores** have diameters greater than 50 micrometre (0.05 mm) and are visible to the naked eye;
- **mesopores** have diameters between 2 and 50 micrometre;
- **micropores** have diameters less than 2 micrometre (0.002 mm)

Soil water is actually held in pore spaces by a suction force which must be overcome if water is to be removed. This force is a result of the natural attraction between water molecules (cohesive force) and between water molecules and soil particle surfaces (adhesive force) which together produce a capillary effect in the smaller pores. For this reason mesopores and micropores may be described also as **capillary pores** and the macropores as **non-capillary pores**. Pores as fine as 0.03 micrometres can hold soil water. In soils containing organic substances and clay minerals there are additional electrostatic and ionic forces increasing the attraction between water and soil. The total suction effect is called the **soil water tension** which is zero in a waterlogged soil but becomes a suction force of many atmospheres value as soil dries out.

Soil water tension

Changes of soil water content can now be re-examined in terms of water tension, pore size and plant requirements. A profile with a range of pore sizes is shown in Fig. 2.1.

Starting with a totally water-saturated, but freely draining soil, all pore spaces are completely filled with water at zero tension and free gases are absent. When rainfall ceases water drains from the macropores from the ground surface downwards under the influence of gravity, leaving air-filled spaces. Since gravitational water is harmful to plants if it remains too long in the rooting zone it may be described as **superfluous water** which must be removed as quickly as possible. When draining is complete the soil is left at field capacity with all soil particles surrounded by a film of water, held at a tension of about 0.3 atm, such that capillary pores remain full of water. This provides an ideal environment for plant roots. Water removed from the film by root hairs at any point is replaced by capillary flow to maintain an even depth of film. Continued plant activity without further

Figure 2.1   Soil moisture tension and pore sizes

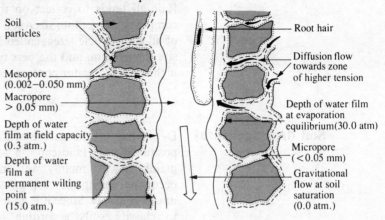

additions of water to the soil creates a soil water deficit as the water film becomes thinner and the tension at which it is held increases. Eventually, at a soil water tension of about 15 atm, plants are unable to remove further water from the film and the wilting point is reached. At this stage the mesopores have largely lost their stored water while the micropores have not. Thus mesopores contain **available water** and the micropores contain water held at such high tensions that it remains in the soil as **unavailable water**. This is not universally true. For example, some plants in arid areas can overcome greater soil water tensions. In many soil profiles the wilting point is delayed because further water may reach the roots from below the rooting zone as a result of **capillary rise** or upwards capillary seepage in response to the tension differential. Finally, in the drying process, at about 30 atm tension, the water film reaches an equilibrium with evaporation which is the limit of air drying.

The effects of pore size on soil fertility

Pore size distribution is of fundamental importance for soil fertility. These characteristics of the various types of soil need to be fully understood. Sands and gravels contain mainly macropores in the inter-grain spaces which allow the rapid drainage of gravitational water but the very small proportion of capillary pores leaves them with little available water to support plants in dry weather. Soils that consist mainly of the finest size of soil particles (clays) have inter-grain pores mainly of micropore size.

In such soils there is little available water and drainage is poor. Rainfall tends to collect on the surface or flow across it. Soils composed of medium sized soil particles like fine sand or a mixture of all soil particle sizes, called loam, have a good balance of pore size distribution and the best types combine good drainage and a high available water content. Such soils can be very fertile.

Soil drainage status

Depending on the nature of the profile, the local climate and its position in the landscape, a soil can exhibit any condition from mainly dry to mainly wet and the condition most commonly occurring during the crop-growing season is described as the **soil drainage status**. Since it is an important factor for crops it is useful to classify soils according to their drainage characteristics. Various systems are used and are based usually on the appearance of soil horizons when observed in a freshly dug pit. The Soil Survey of Scotland has developed a useful system with five grades of drainage status used for mapping purposes. These are: soils with excessive drainage; free drainage; imperfect drainage; poor drainage; very poor drainage.

1 **Excessive drainage**. Usually sandy or gravelly soil profiles with large soil particles and large inter-grain spaces. Drainage of gravitational water is rapid and plants suffer from drought in any dry season.
2 **Free drainage**. Soil profiles which lose gravitational water rapidly but retain adequate amounts of available water to sustain plant growth through most spells of dry weather.
3 **Imperfect drainage**. Soils which lose gravitational water more slowly. If unimproved, waterlogging may persist for two or more weeks after rainfall or the groundwater table may rise into the plant rooting zone long enough to have some effect in reducing plant growth during a wet spell.
4 **Poor drainage**. Soils which lose gravitational water slowly, or are situated where the groundwater table remains high in the profile. In most years the soil root zone loses excess soil water only during the summer months. In an unimproved condition successful cropping is unlikely.
5 **Very poor drainage**. Soils that are waterlogged to the surface for most of the year. Only those species tolerant of waterlogged soil can survive.

It is emphasised that the Soil Survey classification is based on the **natural drainage** qualities of a soil and an area mapped as poorly drained may have land within it which has been improved to a high standard by an efficient land drainage system. The indications of poor natural drainage may persist in a profile long after the drainage condition has been improved.

The plant environment

Variations of soil drainage status are superimposed on a wider pattern of environmental variation caused by the distribution of temperature ranges, rainfall and evapotranspiration values about the surface of the earth. The main climatic zones, known as tropical, subtropical, temperate and arctic, are delimited mainly by temperature ranges. Within each one the distribution of total annual rainfall and the balance of rainfall and evapotranspiration cause further variation in the plant environment. Each kind of environment, in its natural state, is colonised by the plant species that are most suited to it and environments are recognised fairly easily by the distinctive **plant communities** found there. The pattern of vegetation changes can be illustrated by considering the temperate zone. Very dry conditions produce a **desert** where plants are absent. Less harsh conditions with some rainfall allow the growth of cactus and scrub in a **semi-desert**. A minimum of about 250 mm of rain each year allows the development of grasslands which are known as the **prairies** or **steppes**. Where annual rainfall reaches a value of about 500 mm the grasslands give way to **temperate forests**. Within this region the warmer, drier areas are colonised by broad-leaved deciduous trees and the cooler, wetter areas by coniferous trees. All of these are found in the better draining soils. Where wet soils occur the dryland communities are replaced by the characteristic plants of **swamps** and **mires**. Each plant community is in balance with its environment and will not change with time if the environment does not change. Such stable natural plant assemblages are called **climax communities**.

Land use

Most of the temperate forests and prairie grasses have been cleared and replaced by farm crops. Within the forest lands are many areas with soils that are too poor for cultivation and have been left as unimproved grazings. The plants that grow in such areas are naturally established and the communities are in balance with an environment which includes grazing domestic animals. They are, therefore, described as **semi-natural communities**. Intensive grazing eliminates tree seedlings and, in time, the forests disappear. There are a number of distinctive semi-natural communities. In areas with excessively drained soils, cropping is often not worthwhile and a **heathland** community consisting of common heather and other drought-tolerant species provides sparse grazing. On cold, elevated or exposed sites the low temperature reduces the ability of plants to absorb soil water, even when rainfall is adequate, and a **heather moorland** is found. In less severe conditions, where grazing is particularly intensive or where soils are less acid, there is a tendency for species of grass to replace the low shrubs and a range of **grass moorland** communities develop with variations of soil acidity and drainage status indicated by the grass species present. Large tracts of the better grass moorlands have become invaded and dominated by the bracken fern. Where soil drainage is poor all types merge into the semi-natural communities of the **wetlands**. These have distinctive species of interest to land drainers. In very poorly drained sites true climax communities of swamps and mires are found, often undisturbed.

The effects of poor soil drainage on agriculture

The effects of poor soil drainage on farming operations and the general symptoms of restricted drainage vary from site to site depending on the severity of the problem. It is the task of the land improver to recognise the symptoms and to evaluate the loss of income compared to the cost of improvement. It is convenient to discuss these in terms of the severity of the drainage problem.

The wettest sites

Those sites with the poorest drainage status will usually have a natural vegetation and agricultural use, if any, is limited to extensive grazing. The symptoms of restricted drainage will be clear and the characteristic communities of each type easily recognised.

- **A standing water environment** – in standing water a range of floating and submerged or partially submerged species is found. The group includes pondweeds, duckweeds, water crowfoots, water buttercups, mairstail, water violet, arrowhead and water lily.

- **A water margin environment** – along freshwater shore-lines where sediments are rich in plant nutrients a **reedswamp** community can flourish. The most common species is common reed which may form a pure stand, but elsewhere it is mixed with or replaced by similar species like sedge, bur reed, reed-mace, bulrush, fen sedge and many others.
- **A wetland environment** – on sites where there is no surface water. The climax community of any wetland in a warm dry climate is **carr** woodland where reedswamp species give way to willow and alder scrub.
- **A peat mire environment** – in regions which are cool and wet the climax vegetation is a **peatland** community. The dominant species are sphagnum moss and cotton sedge. Other plants commonly found are cross-leaved heath, tufted scirpus and bog asphodel while common heather occurs on the drier hummocks.

All such sites are easily recognised and most wetland sites have a cover of peat.

Moderately wet sites

Sites with poor drainage are widespread in high rainfall areas and elsewhere where drainage is restricted by topography. Land use depends on circumstances and ranges from rough grazings to arable cropping. Most unimproved grazings consist of semi-natural communities containing the characteristic species of poorly drained soils. These included rushes, bog bean, marsh thistle, creeping buttercup, marsh St John's wort, purple moor grass and tufted hair grass. Attempts to introduce better species of grass without drainage improvement are likely to fail because they cannot compete with the native species. Such sites are also fairly easy to identify.

Better sites

In more productive farming areas the climate of a particular year, or of a season when crops are at a sensitive stage, may have more influence on the success of a crop than the drainage status. In very dry years soils that might be classified as poorly drained can produce successful crops while in very wet years there can be crop failure in land considered to be freely draining. In general, however, the expectations of successful crops will be reflected in the rotations followed and by choice of crops. Soils with drainage restrictions in an average year will normally be left as permanent pasture while arable cropping will increase in soils with better drainage. Arable cropping is generally possible in imperfectly drained soils but very sensitive crops like peas or difficult-to-harvest root crops will be avoided. Drainage limitations are less obvious in such circumstances.

The general effects of restricted soil drainage

Arable crops

Arable crops can suffer in various ways. Soils with drainage problems take longer to dry out in spring so the sowing time is delayed or the seed is placed in a wet, anaerobic seedbed. The seed may fail to germinate or it may die soon afterwards. If waterlogging occurs after germination the young plants may die out. All of those conditions cause bare patches where the wettest areas occur in a field. The crop plants which do survive will always be retarded. When a growing root tip encounters a saturated zone in the soil, growth in a downward direction stops and branching occurs in the aerobic zone above. When the zone of saturation rises into the rooting zone the existing roots die. In either case effective roots are restricted to soil near the surface, rendering the plant more susceptible to drought if the soil dries out in any later dry period. Other difficulties are nitrogen deficiency due to losses of nitrates and the development of toxic substances, both caused by anaerobic conditions. The resulting crop, therefore, is patchy with bare land surrounded by zones of sparse, pale coloured plants merging into areas of healthy plants where the land is drier. The bare patches and thinly populated areas are soon occupied by weeds which are typical of wet soil in cultivated fields and include redshank and creeping buttercup, but in dry periods most of the other arable land weeds can become established. A further yield reducing factor is the suitable environment in wet soils for plant diseases like take-all and club root.

Grassland plants

Grass plants are retarded in the same way as arable crops with roots restricted to a shallow zone above the level of profile saturation. If the topmost soil horizon is compacted by grazing animals, the damaged horizon can dry into a cement-like crust just at the time when the zone of saturation begins to move downwards away from the shallow roots. The hard soil prevents root growth, causing the plants to suffer from drought despite the wetter subsoil, and all leaf growth stops. The loss of nitrates and accumulation of toxic chemicals occur in grassland with loss of yield just as in arable crops. Poor conditions for cultivated grasses allow entry of less productive wild species which gradually eliminate the sown species. Shorter growing seasons and lower yields of grass are not the only problems. Animal production also is limited by the irritation of flies and the occurrence of diseases like foot rot, liver fluke and parasitic worms, all of which find wet pastures an ideal environment.

Soils

Soils which contain appreciable amounts of clay particles are plastic when wet. In this conditions the soil is easily moulded and when subjected to field traffic such as cultivating and harvesting machinery or grazing animals, particularly cattle, the surface can become compacted, smeared or rutted. In the worst conditions the surface becomes churned into near liquid mud and machine wheels may sink deeply into the profile. Cattle walking on a soft surface produce a layer of mud which buries the grass. This is called **soil poaching** and is particularly prevalent in young leys since old pasture with a thick matted turf is more resilient. Implements pulled through wet soil, together with the effect of spinning tractor wheels, produce a common type of compacted layer, just below the depth of cultivation in arable fields, called a **plough pan**. All slurried or compacted horizons can dry out to become hard and impervious resulting in the creation of a barrier to both air and plant roots. Drainage is impeded, thus forming a self-perpetuating deterioration of soil drainage status and fertility. Sandy soils do not become plastic when wet and are less easily damaged.

The need for land drainage

The benefits to be derived from an improved drainage status can scarcely be doubted. All improvement projects must be preceded by examination of the site to determine the drainage status and to identify the cause of any poor drainage status so discovered. Thereafter a management decision is required to assess the likely costs of the improvement and the possible benefits to be derived from it. There are many variables in the equation but the most important are the capital costs of the project, the running and maintenance costs and the amount of additional income. All of these factors change from site to site and from time to time. Market returns are particularly variable.

Assessing the soil drainage status

Identifying the extent and nature of unsatisfactory soil drainage is an essential first step. In most cases much evidence is available and all of it should be considered before making a decision.

Cropping records

Histories of crop failures and successes related to climate variations should be considered together with limitations of choice of crops compared with other fields. Performances should be compared with those of similar soils nearby whenever possible. There will be cases of deteriorating drainage status where crops successful in the past are no longer worthwhile.

Surface conditions

The most obvious signs of poor drainage will be seen on the surface of the ground. Walking over the site may reveal soft patches, springs, standing water, ruts or tracks, but often the evidence is less clear cut. In springtime the different rates of soil drying may reveal wet areas. The wet areas may relate to low-lying parts of the field, old drainage systems or to changes of soil types. In some cases they may have no obvious relationship with other factors. In summer the crops may have bare patches or areas of unhealthy plants or the soil may have hardened into a resistant crust which sometimes cracks into rough polygons. In autumn and winter the absence of stubble in patches, or sparse stubble or obvious signs of difficult harvesting give a strong indication of crop success.

Vegetation

The identification of plants which thrive in wet conditions is a useful diagnostic feature. On grassland the most common indicators of wet soils are the various species of rushes, creeping buttercup and grasses like purple moor grass. In arable land the appearance of crop plants and the presence of wet soil weeds like buttercup and redshank are useful guides. Where natural or semi-natural communities are present there will be no difficulty in identifying wetland sites. For more comprehensive guidance on plant communities and for details of local plant species it is recommended that reference be made to standard texts on plant communities and ecology.

Soil profiles

The most import information on soil fertility and drainage status is obtained by digging a pit and examining in detail the profile properties. This is discussed in Chapter 4.

Existing drainage systems

Examination of the existing drainage system will be valuable. Nearby waterways and fieldside ditches are necessary to carry away the drainage water and they need to be in good condition. If they are full of standing water then the groundwater table cannot be at a lower level in the field. The outlets of field drains may have become blocked. Examination of field drains may provide evidence of blockage or restriction of gravitational water movement towards the drains. All such evidence is important and indicates whether better maintenance, better soil management or new drainage works are needed.

The need for conservation of wildlife

It is technically feasible to clear, drain and cultivate nearly every type of plant environment. There remains, however, the need to consider the desirability of preserving some minimum amount of the natural environment for the preservation of all types of wild species. In many countries the most valuable natural habitats have been identified already and have been given a measure of protection by legislation. These protected areas are called Nature Reserves, National Parks, Sites of Scientific Interest or Areas of Outstanding Beauty. But they are not protected absolutely and many other unlisted sites need to be protected from change since they add variety to the habitat. Generally, the best farming lands should be drained and farmed to achieve maximum production using methods that do not impair the long-term fertility. The sites to be left undisturbed are the woods, moors and wetlands which would only be marginally economic, if at all, after improvement. There is no point in destroying a natural habitat that cannot be converted into economic farm land. Furthermore, there is a case for setting aside some of the better land, particularly the hedgerows and field corners. Each land occupier should take the responsibility of balancing food production with conservation. It may be worthwhile seeking the advice of a wildlife specialist.

Chapter 3 Geological processes affecting land drainage

Several factors of importance for natural soil drainage are influenced directly or indirectly by one or more geological processes. These include soil formation, the characteristics of flowing groundwater through different rock types and the nature of land forms. For many drainage problem sites it is helpful to consult geological maps and for this reason alone it is worthwhile to become familiar with geological terms.

Some definitions
Cycles of erosion

Every rock mass exposed at the surface is subjected continuously to the processes of rock decay and, in time, nearly every part of the land surface is covered by a layer of loose fragments. This loose material tends to be removed from its source area, ultimately to be deposited in the sea and the rate of removal is proportional to elevation above sea-level. Constant removal of loose material reduces land surfaces to near sea-level, at which point the rate of loss is much reduced. The earth's crust, however, is not static and from time to time areas of crust are elevated once more to begin a new **cycle of erosion**.

Rock formations

The term **rock formation** describes all forms of distinct layers and masses that make up the crustal shell of the earth. It includes the still unconsolidated assemblages of loose material like gravel mounds and lake bottom muds. There are many types of rocks but they may be grouped into three basic types: **sedimentary**, **igneous** and **metamorphic**.

Rock structures

Rocks are not homogenous masses. The fabric is composed of grains or crystals of various shapes and sizes fitting together with or without intergrain spaces. Many rocks are found to be fractured. Nearly all hard rocks have sets of parting planes often trending at right angles to each other. The two sets of vertical fractures are likely to be the result of crustal flexing while the horizontal set is the result of differential expansion resulting from release of confining pressure as the rocks above are eroded away. These fractures are called **joints**. On a larger scale compressive forces on crustal rocks can throw the layers into **folds** or may move one mass of rock, relative to an adjacent mass, along a plane of movement called a **fault**, if the movement is relatively vertical, or a **thrust**, if movement is relatively horizontal. All such features are described as **rock structures**.

Rock permeability

Rocks with interconnecting macropores in the form of intergrain voids or fractures which allow groundwater to pass through the rock mass are described as being **permeable** to groundwater. Formations that are tightly compressed or composed of fine grains and are not fractured are **impermeable** to groundwater. The terms **pervious** and **impervious** are sometimes used but have the same meaning. Joints and faults have a marked effect on the permeability of rocks near the surface where release of pressure has allowed all fractures to open up, creating easy pathways for groundwater movement.

The characteristics of rocks
Sedimentary rocks

Sedimentary rocks form on the sea bed where rivers deposit rock waste removed from the land surface. Flowing water grades the rock waste according to particle size such that there is a tendency for coarse sands to collect near the river mouth, fine sands build up further out and clay particles are spread over a wide area. The sediments become deeper as time passes and seasonal or other changes of deposition rates cause a distinct layering with separate layers often clearly defined by **bedding planes**. Major changes in the environment may alter the respective areas of accumulation causing, for example, clay particles to be deposited on top of sand or sand on clay. Other events in and around the sea can produce layers of calcium carbonate, salt or plant remains and eventually a **sedimentary succession** is formed. As time passes new layers bury old ones more deeply. Deep burial causes intense pressure, compacting loose grains into hard rocks. In this way sand grains become coherent **sandstone**, clays or muds are compressed into shale and fragments of calcium carbonate become **chalk** or **limestone**. Crustal movements have elevated many of these sedimentary rocks which now form the land surface in many parts of the world. Sedimentary successions exposed after a relatively short period reappear as sand or clay and form landscapes of low relief as seen in south-east England or northern France. Successions elevated after longer burial are harder and often folded and faulted and provide landscapes with more pronounced features like the Pennines in England or the Jura in France. Sands, sandstones, chalk and some limestones usually are permeable because of intergrain spaces. Clays, shales and the more massive limestones usually are not permeable. Mixed deposits usually have intermediate permeability values and any rock with secondary materials deposited in rock voids, like cemented sandstones, may be impermeable. Most formations of hard rocks are rendered permeable near the surface because of open joints.

Igneous rocks

Crustal rocks may be invaded by a high temperature, fluid rock material called **magma**. The magma then cools in a variety of conditions and solidifies to form igneous rock. Large bodies of magma cooling slowly at depth produce coarsely crystalline rocks, the most common of which is granite. Rising magma can invade fault lines through other rocks where it cools to form a vertical sheet called a **dyke** or it may force its way along the bedding planes to form a horizontal **sill**. Should the magma reach the surface it may erupt as a shower of ash or it may flow over the surface where it cools to form **lava**. Dykes, sills and lava are composed of much finer crystals than granite and have the common name of **whinstones**. Most igneous rocks consist of closely inter-locking crystals and have no pore spaces. Uncompacted ash may be permeable but ancient layers are usually impermeable. Some types of lava contract on cooling – notably basalt – and split into columns. These and jointed igneous rocks can be highly permeable.

Metamorphic rocks

Major crustal disturbance can cause severe compression and buckling of the formations involved. When pressure and temperature are extreme, a partial fusion and re-crystallisation of the rock-forming materials may occur, resulting in the occurrence of metamorphic rocks. Sedimentary, igneous and existing metamorphic rocks can all be altered to form new rocks with characteristics that reflect the conditions in which they were formed. Originally permeable rocks can lose pore spaces and whole sedimentary successions may be obliterated with distinct members of the succession being recognisable only by close inspection of the minerals present. When exposed, metamorphic rocks are hard, brittle, crystalline formations which are much fractured and permeable to groundwater.

Rock weathering

Most rocks and rock-forming minerals evolve in conditions of high temperature and pressure deep below the surface where volume is reduced to a minimum and mobile material like water is forced to migrate upwards. When these rocks come to be exposed at the surface they decay easily because both physically and chemically they are unstable in this environment. The many processes which cause the decay of rocks are known collectively as **weathering**. Rock weathering is really another kind of metamorphism which forms new materials that are stable at the surface, a process which requires an increase in volume and the addition of water. There are two main groups of decay processes which work together, each enhancing the effectiveness of the other.

Physical weathering

Physical weathering or rock disintegration is the total effect of a number of forces. They disrupt rock fabric from the ground surface downwards until the layer of loose materials is thick enough to protect the underlying rocks from further action. Thereafter the rate of disintegration keeps pace with loss from the surface. The more important agencies of destruction are differential expansion at the surface as pressure of overburden is released, heating and cooling cycles which are most effective in hot desert areas and freezing and thawing cycles affecting groundwater. The expansion of interstitial water, when it freezes, is the most destructive force and physical weathering is most active in the northern margins of temperate regions where freezing and thawing cycles occur most frequently.

Chemical weathering

Chemical weathering or rock decomposition results from a number of chemical processes and from biological activity. The end result is the formation of new substances with greater volume. These new minerals lack the strong interlocking connections of the original crystals causing rocks to crumble, but not necessarily from the surface downwards, as with physical processes, since chemical reactions depend on the presence of groundwater as a medium of chemical activity. The rate of decomposition depends on the features described below:

- *The porosity of the rocks*. The more pore spaces there are the greater is the area of interface between rocks and water. Porous or jointed formations can decay more rapidly than those that are impermeable.
- *The chemistry of the rocks*. Rock-forming minerals are not uniformly susceptible to decomposition. The constituent particles of sedimentary rocks have been altered, at least partly, in an earlier cycle or erosion and cannot be further altered to any great extent. Crystalline igneous and metamorphic rocks resist alteration to a degree that depends on their constituent minerals. In general, light-coloured rocks like granite are more resistant than dark-coloured rocks like basalt.
- *The presence of acid groundwater*. Carbonates from decaying plant litter and sulphates and nitrates from the atmosphere can convert groundwater into a weak acid which attacks rock minerals.
- *The retention of groundwater*. Rocks that are continually saturated by groundwater or rainwater are more rapidly decomposed than those that are relatively dry.
- *The temperature of groundwater*. The rate of chemical reactions tends to increase with temperature so that rock decomposition is most active at the equator and least active at the poles.
- *Biological activity*. Soil micro-organisms play an active part in the final breakdown of minerals to release plant nutrients. They are most active in warm soils rich in organic materials.

Rock waste Weathering maintains a layer of **rock waste** on top of the solid rocks that varies in thickness from a few centimetres to tens of metres depending on environmental conditions. In some areas it is absent due to recent severe surface conditions. Physical weathering alone produces fragments of rock with unaltered minerals and should the rock waste remain in this condition to form new sedimentary rocks the sediments are said to be **immature**.

Chemical weathering creates new substances which do not resemble the original minerals and any sediments so formed are said to be **mature**. Generally, in all except the coldest environments, chemical weathering is sufficiently active to cause some decomposition. Where it is most active, in equatorial rain forests, all rocks have been converted to a loose, soft mass to a depth of at least 30 metres. Rock waste situated where it was formed is a **sedentary** formation which reflects the characteristics of underlying rocks, as would be expected, with sandstone becoming sand and shale becoming clay. Rock waste may be moved from its source area by the processes of **erosion**, **transportation** and **deposition**. Many areas have rock waste that has been transported and deposited on top of local rocks so that the rock waste does not necessarily resemble the rocks below. All such transported surface layers are called **drift** and have properties influenced by the nature of the transporting agency. The forces of erosion, transportation and deposition are flowing water (channel flow and surface flow), moving ice (ice sheets and valley glaciers) and flowing air (wind).

Water-transported drifts

Drifts resulting from water movements are usually of limited extent and are normally restricted to river valleys and coastal plains. Drifts associated with river valleys are discussed more fully in Chapter 8. In arid lands an important feature is an **alluvial fan** which forms where storm water from a hilly area reaches a plain and deposits the transported sediments. Alluvial fans are mainly permeable. The horizontal formations of marine sand and clay left by a retreating shallow sea are not strictly drift but have similar properties. In formerly glaciated lands these often narrow, coastal plains are called **raised beaches** and have a permeability that depends on constituent particle size. Horizontal, uniform surface deposits of marine origin are of limited extent in world terms but are widespread in the Netherlands where land drainage was pioneered. Such uniform layers are easily reproduced in a laboratory and, as a result, have had great influence on the evolution of land drainage theories.

Ice-transported drifts

Most of the drift found in the northern temperate zone can be associated with various phases of the Ice Age. Marked climatic fluctuations caused at least four major periods when ice spread well beyond its present limits. Episodes of ice advance or glaciations were separated by interglacial periods of mild weather lasting up to about 60 000 years each and the most recent glaciation ended about 10 000 years ago. During glaciations precipitation occurred as snow in present high rainfall areas of temperate climate, particularly in the mountains of north-west Europe and the northern parts of North America. Deepening ice on high ground flowed outwards radially, often across lowlands and areas of lower precipitation until it reached a point where the rate of ice melt equalled the rate of ice flow. This limit of ice cover was rarely static and changed with variations of temperature and precipitation.

The effects of advancing ice

At high elevations and in polar regions ice sheets remain frozen to the underlying rocks so that ice flow must occur over shear surfaces within the ice and very little erosion results. But in valleys or temperate lowlands the pressure of overlying ice melts the layer in contact with the ground. The moving ice sheet scrapes over the land surface and, since rock fragments become embedded in the ice, erosion can be severe. The type of surface left by the ice depends on the gradient and the proximity of centres of ice accumulation.

- Near to the centres of ice dispersal, especially where the ice had a short downslope route to the sea, denudation is total, leaving a surface of bare, often scraped or polished rock. Such features occur in the fjorded coastlands like those of Norway, Scotland and British Columbia.
- Where the ice had a longer overland route the rock waste moved by the ice so increased in volume that it tended to protect the surface from further erosion and resulted in some deposition which increased in amount towards the limits of ice advance. The ice tended to roll the loose material forwards, causing mixing and grinding, finally depositing a non-layered drift of often fine grained material randomly mixed with various sizes of stones and boulders. Drift originating under the ice is variously called **ground moraine**, **boulder clay** or **glacial till**. It was compressed into position in a saturated state (that is, it was puddled), producing what is now found to be a usually compacted drift that is impermeable to groundwater.

It can vary in thickness from a few centimetres to hundreds of metres where a cross valley in the line of ice flow has been filled, but most of it is less than 10 metres deep. Drift of this type is found in Scandinavia, the British Isles north of the River Thames, the northern fringe of Europe from Holland through to northern Russia, most of Canada and the northern fringe of the United States.

The limits of ice advance and minor re-advance episodes of the last retreat are often marked by more or less conspicuous, irregular, linear mounds of rock waste called **terminal moraines** which are composed of all rock fragments carried along by the ice. They are mainly gravel and stone deposits of little value to agriculture. The best examples are seen in southern Finland. Terminal moraines are sometimes associated with irregular mounds of local rock waste moved along the front of the ice and called **push moraines**.

The effects of retreating ice

A different set of drift deposits is associated with the retreat of the ice sheets. When a major climatic improvement occurred the ice did not merely retreat but began to melt over its entire surface causing elevated parts of the landscape to appear above the ice, leaving stagnant ice in valleys and lowlands. Vast quantities of melt water cut channels and gullies through ice and landscape alike and flowed in sheets across newly exposed plains. Rock waste was variously eroded, collected, sorted, deposited or carried away as flow rates fluctuated. Fine grained materials carried by the ice were usually swept away to sea but coarser fragments were left behind in mounds of various shapes resting on glacial till and collectively called **fluvioglacial drift**. All such drift is composed of layered deposits of sand, gravel and stones most of which are excessively drained and many are of limited value for agriculture. Some sand was carried beyond the limits of glaciation and deposition as **outwash drift**. This also is infertile, excessively drained material which is extensive in the heathlands of north Germany and Poland. Exposed glacial till could be eroded by flood water or its surface could be reworked by an episode of severe weather and rendered more permeable.

The effects of severe weather beyond ice sheet cover

Beyond the limit of ice cover at any stage of ice advance the severe climate caused a tundra environment with permanently frozen ground. The summer melt produced a waterlogged slurry above a still frozen subsoil. On sloping sites the slurry could flow downslope year by year until hilltop sites were stripped of all rock waste and **solifluxion drift** collected at the bottom of slopes. This resembles glacial till in that it is compacted and usually of limited permeability but it is layered with any contained stones lying with their longest axes parallel to the bedding planes. A named example is the Coombe Rock at the bottom of chalk hills in England. On more level sites the summer slurry could dry out and polygonal patterns of cracking could become sites of **frost wedges** which could reach a depth of several metres. When the ice melted the cavity became filled with loose surface material. Frost heave could fill the cracks with stones or coarser material and the squeeze of re-freezing from the surface could cause mud flows. All such effects persist today and are seen as **patterned ground**. In springtime when cultivated soils dry out they can be seen as rough polygons or parallel, wavy lines which mark the contrasting zones of soil particle sizes. All have the effect of causing variability of soil properties across the field.

Wind-transported drifts

Winds crossing dry, unprotected surfaces can move loose particles to a greater or lesser extent depending on wind speed and particle size. Coarser particles like sand grains can be moved in a series of hops and build up against any barrier which may itself be composed of sand. A feature of any sandy area such as a desert or sandy seashore is the slow downwind advance of **sand dunes**. A spring 'blow' is quite common in the drier areas of sandy or peaty soils when strong winds can move topsoil, together with seeds and fertilisers, into nearby drainage channels or onto roadways. Such land needs protection to safeguard the topsoil until crops can become established. Winds blowing across desert areas, whilst moving sand grains, winnow out silt-sized particles which are carried a long way, often returning to the surface only when rain is encountered. Any moist region downwind from a desert, therefore, will receive a constant supply of silt grains which accumulate as a layer of drift. Such drifts are widespread in western China and near desert areas in the central states of North America, particularly Kansas and Nebraska. The drift is a buff-coloured, homogenous layer of silt-sized grains which is called **loess**. In wetter regions of Europe and North America loess has developed for another reason. Glacial erosion produced large quantities of finely ground rock waste which dried out in the cold desert conditions at the margins of an ice sheet. Winds blowing off the ice sheet transported these particles into wetter areas where loess is now found. Although much fragmented by later erosion, such deposits are widespread from northern France to eastern Europe where a near continuous cover merges with the desert loess of central Russia and China. In a similar way the central plains of North America are loess-covered as a result of glaciation and there also it merges with desert loess. Loess produces a soil that is only moderately permeable.

Soil-forming processes

Rock waste is the raw material on which soils develop. The surface layers are colonised by a great variety of life forms which accelerate chemical decomposition and add organic waste products. In time the raw rock waste is converted into a fertile soil. Since nearly all soil-forming processes are influenced by the environment, there is a strong correlation between climate and soil characteristics in a region. For land-drainage purposes it is worth considering the sequence of events that convert a crystalline rock into a mature soil and for this purpose the widely occurring granite is most suitable.

The composition of granite

Close examination of granite reveals that it consists mainly of grey or pink, porcelain-like crystals of feldspar which can have flat surfaces and give the rock its characteristic colour. Interlocking with the feldspar crystals are irregular, dark, glassy crystals of quartz which account for about one third of the rock volume. More or less evenly dispersed amongst the larger crystals are shiny metallic or shiny black flakes or speckles of two types of mica. These three very common minerals of crustal rocks are largely responsible for the nature of rock waste and their weathered derivatives make up its bulk.

Weathered granite

Physical weathering produces a shallow layer of fragments consisting of unaltered crystals of feldspar, quartz and mica. Chemical weathering converts feldspar into a soft mineral called kaolinite which is a kind of clay. The mica crystals are converted into other types of clay minerals. Quartz crystals, however, totally resist chemical change and survive intact. Complete decomposition results in a mass of clay with quartz grains embedded in it.

However, the rock fabric is destroyed at an early stage of chemical weathering when only a thin layer of each feldspar crystal has been altered. At this stage the rock mass consists of a loose, crumbly aggregate of mainly fresh crystals; such a sedentary formation is called **grus**.

Depth of weathering

In a tropical climate granite is weathered to great depth. Groundwater fills all the joints and attacks each block on all of its surfaces. The decay process advances towards the centre of each block and, as a result, the rate of decay is influenced by the amount of jointing present in the rock. At some depth below the surface, decomposition will not have reached the centre of the blocks and a rounded **core stone** of still fresh granite will be found in each one resting in a mass of grus. Deep chemical weathering occurred in the present temperate regions during the much warmer pre-glacial times and in such regions the bare rock surfaces now to be observed may represent a loss of 30 or more metres depth of rock waste. Elsewhere glacial erosion has converted the grus into glacial till containing core stones, solifluxion has stripped all loose material from elevated sites leaving piles of core stones called tors or water erosion has dispersed the decayed granite and sorted it out according to particle size before deposition in water as sediments. In the less severely eroded areas undisturbed grus is often seen lying below glacial till where it merges downwards with still fresh granite.

Plant nutrients

Other crystalline rocks are composed of different sets of minerals when compared to granite and are altered in the same manner. Quartz always remains unchanged and subsequent attrition may break the particles in smaller sizes which survive as grains of **sand** or **silt**. Most other minerals are altered to the many types of clay minerals and metallic ions may pass into solution in groundwater or become attached to clay particles. Feldspar crystals are more or less universal and different types release potassium, calcium and sodium. Sodium is quickly lost from the rooting zone. Most igneous rocks contain occasional crystals of apatite, which is the primary source of phosphorus. Nitrates are obtained from the atmosphere.

Clay minerals

The term **clay** is used to describe the smallest particle sizes found in soils, but in a geological context it represents a large number of minerals. Clay minerals have many common properties. Most clays are strongly cohesive and form a plastic material when wet but can be rendered hard and resistant by drying and heating. However, the different minerals can have distinctive properties which affect the nature of the soils in which they occur. Clay minerals can be arranged in three groups according to their physical and chemical properties. The **kaolinite** group has crystals with a structure of atoms which permits molecules of water and metallic ions to become attached only to the outer surface. The plasticity and cohesive properties are well developed but shrinking and swelling properties are relatively weak. The **smectite** group has a different atomic structure with a looser bonding of constituent sheets which gives the crystal chemically active inner surfaces. As a result they tend to shrink and swell more conspicuously than the other groups. The **micaceous** clay group contains minerals which are much less cohesive and tend to convert easily into liquid slurries when wetted. They have little tendency to shrink or swell. Kaolinite sediments are common in rocks of carboniferous age, smectites are often the dominant type in Jurassic and Cretaceous sediments, while micaceous clays often result from the glacial erosion of rocks that are rich in mica, particularly ancient metamorphic rocks.

The effects of geological processes

Soil properties are the result of the kinds of rock waste produced by geological processes and the effect of biological activity over a long period. The different climatic environments give rise to distinctive features in soil horizons which are used as a basis of soil classification. These are discussed in Chapter 4. The general arrangement of rock formation and the relationship between permeable and impermeable layers in rocks near the surface have considerable influence on groundwater movements and are considered in more detail in Chapter 9. Different rates of surface erosion directly influence the shape of a landscape and this important aspect of drainage design is reviewed in Chapter 6.

The use of maps

The different rock formations in an area can be identified by consulting the geological maps that are available in many countries. They show clearly the limit of outcrop of the different type of rocks and identify important features like major faults and the angle of dip of layered rocks from the surface. Most maps include an identification of drift with each type marked as an overlay of symbols on the colours which identify the rocks. Many of the more important agricultural areas have now been mapped to show the extent of the different types of soil and these identify the drift and sedentary rock wastes that are the soil parent materials.

Chapter 4 **Mineral soils**

The diagnostic features of the soil physical properties and drainage status can be used, in many cases, as a means of identifying drainage requirements.

Soil profiles Simple observations indicate that soils consist of varying proportions of:

- Solid inorganic materials like stones, sand and clay.
- Organic matter like living and dead roots, various animals and their products, microbiological populations and organic materials in all stages of decay.
- Pore spaces and fractures between the parts of the solid framework which are occupied by water and air in varying quantities.

The way these constituents are linked together, the relative proportions and the different form of each can vary markedly in both lateral (across the field) and vertical (from the surface downwards) dimensions and also with time. Agricultural productivity, drainage problems and their solution, cultivation and harvesting techniques and other factors are closely associated with the properties of the soil profile in a field. Close attention to soil profile characteristics is essential to get the best results from the land. Soil variability is determined by the nature of the rock waste. In the great plains of the continental interiors the loess forms a remarkably uniform soil that varies little over hundreds of kilometres while, at the other extreme, glaciated lowlands in maritime regions have a drift cover noted for its variability. The resulting soils are variable within a few metres and in order to assess the overall soil properties of an area it is necessary to examine a number of representative profiles at different sites within the area.

Soil profiles exhibit many different properties and some of these may be discussed in the following categories:

The characteristics of soil profiles

- *Chemical*. pH, lime requirements, available and total nutrients, gases in the soil atmosphere, mineral and organic ions dissolved in soil solution.
- *Physical*. Texture, structure, depth, permeability, colour, stoniness, water content, bulk density, hydraulic conductivity.
- *Organic*. Living and dead biological material, roots, bacteria, humus.

It is the physical properties which are of greatest importance when considering soil drainage problems. It is rare for two soil profiles to be absolutely identical. Many of these properties, however, can only be determined accurately in a laboratory using sophisticated equipment so they cannot be considered in the field. Bearing these points in mind and recognising that in certain cases it may be necessary to make more detailed laboratory or field tests, the most useful information for drainage purposes may be obtained by the field examination of horizons, depth, texture, structure, permeability, stoniness, colour and organic matter.

Soil horizons

Most soils have more than two distinct soil horizons and generally each soil has several horizons. A number of broad horizons are recognised and, for pedological or mapping purposes, these are then subdivided and described in detail.

- The **A horizon** or topsoil is the upper layer and usually is darker in colour than other horizons because it contains more organic matter. In natural (undisturbed) profiles it can be divided into minor discrete horizons (A_{00}, A_0, A_1, A_2 and so on) but in cultivated soils the A horizon is thoroughly mixed and referred to as the plough layer (Ap).
- The **B horizon** or subsoil underlies the A horizon and generally contains more clay, has less organic matter and is lighter in colour. This horizon also exhibits subdivisions and these are mostly undisturbed by cultivations. Taken together the A and B horizons are referred to as the **solum** or true soil.
- The **C horizon** is commonly called the **soil parent material** and is rock waste.

Soil texture

Soil texture refers to the size of individual mineral particles (<2 mm in diameter) and to the proportions of the various grades of particle size present in the soil. The different grades of size are determined by particle diameters or equivalent diameters (if not spherical) and the terms sand, silt and clay are used. Sand grains of all grades and silt particles are usually different sizes of fragments of quartz crystals while clay particles consist of various clay minerals and metallic compounds down to molecule size. A number of systems are used to define particle size and a satisfactory scale is shown in Table 4.1.

Table 4.1   Soil particle sizes (mm diameter)

Stones and gravel	> 2.00
Coarse sand	2.00–0.2
Sand	0.2 –0.1
Fine sand	0.1 –0.06
Very fine sand	0.6 –0.02
Silt	0.02–0.002
Clay	< 0.002

It is instructive to visualise the relative sizes of soil particles. It will be noted that silt grains are ten times bigger than clay particles, sand grains are ten times bigger than silt and coarse sand grains are ten times bigger than sand. Many soil properties, including drainage, depend on the proportions of the three mineral fractions present – sand, silt and clay – each of which is referred to as a **soil separate**. Fragments larger than 2 mm are described as **coarse earth** and include gravel, pebbles, stones and boulders. Soils can be classified according to the proportions of soil separates present and the names used are self explanatory except for the use of the term **loam** which is used to describe a mixture of separates such that the soil has intermediate properties. For reference purposes the main texture classes with conventional symbols are: sand (S); silt (Z); clay (C); loam (L); loamy sand (LS); sandy loam (SL); silty loam (ZL); clay loam (CL); sandy clay loam (SCL); silty clay loam (ZCL); sandy clay (SC); and silty clay (ZC).

Assessing soil texture

Soil textures can be established in a laboratory by **mechanical analysis** which is based on the rate of sedimentation of the different fractions. Alternative standard methods using hydrometer, pipette or centrifugation are available. Organic substances and coarse earths are first removed and soil aggregates are broken down before the texture (per cent sand, per cent silt, per cent clay) of the sample is measured. The soil can then be classified using a texture chart as shown in Fig. 4.1.

Soil texture can be assessed by hand. With experience and care it is possible to assess texture in the field fairly accurately (compared with mechanical analysis) and reproducibly (compared with other observers). The method is to take a small sample of

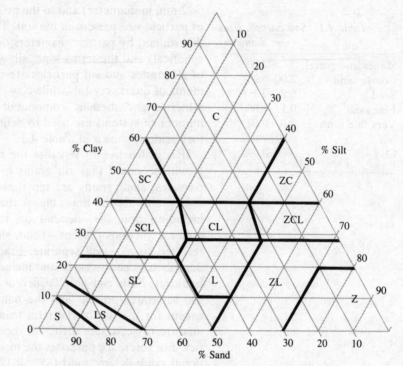

Figure 4.1 Soil texture chart

soil and remove any coarse earth and roots present, then moisten the soil and rub it between the fingers until it is just wet enough to mould. The 'feel' of the soil depends on the texture.

- *Sand*. Feels gritty or hard and lacks cohesion (that is, the particles do not stick together). It is loose when dry and not sticky when wet.
- *Silt*. Feels soft and smooth like talcum powder.
- *Clay*. Feels sticky, is cohesive and can be moulded into a ring shape (that is, it does not fracture).
- *Loam*. Is readily moulded into a ball with particles binding well together. Although sand is present the soil is not obviously gritty, the quantity of silt is not sufficient to give the soil a silky feel nor is there enough clay to make it sticky.

It is necessary to make allowance for greater than normal amounts of organic material which tends to impart a silky feel (texture refers only to the inorganic constituents). Organic matter also increases the cohesion of sandy soils and reduces the cohesion of clay soils. A logical sequence for texture assessment by hand is shown in the key in Table 4.2.

Table 4.2   Assessment of soil texture (*mineral soils*)

1. Rub the soil between the thumb and forefinger, attempting to assess its predominant feel – if this is:

    (a) Gritty   —— 2

    (b) Sticky   —— 5

    (c) Silky or soapy   —— 8

    (d) Neither (a) nor (b) nor (c), forming a fairly cohesive ball   —— **Loam**

2. The soil is predominantly gritty, does not bind well and:

    (a) does *not* form a cohesive ball   —— 3

    (b) does form a cohesive ball   —— 4

3. The soil is predominantly gritty, does not bind well and:

    (a) does *not* stain the fingers   —— **Sand**

    (b) does stain the fingers slightly   —— **Loamy sand**

4. The soil is predominantly gritty, binds well into a ball and

    (a) breaks up when moderate pressure is applied   —— **Sandy loam**

    (b) feels significantly 'sticky' but takes a polish only with difficulty   —— **Sandy clay loam**

5. The soil feels sticky and

    (a) is quite gritty in feel, takes a polish and deforms only with difficulty   —— **Sandy clay**

    (b) feels significantly soapy or silky and takes a polish   —— 6

    (c) is neither gritty nor silky, holds together strongly and takes a polish   —— 7

6. The soil is predominantly sticky and takes a polish, a silky feel is readily discernible and

    (a) it is moderately difficult to deform   —— **Silty clay loam**

    (b) it is very difficult to deform   —— **Silty clay**

Table 4.2 (con't)

7. The soil is predominantly sticky, feels neither gritty nor silky, forms a cohesive ball and takes a polish, while:		
(a) it is moderately difficult to deform	——	**Clay loam**
(b) it is very difficult to deform	——	**Clay**
8. The soil feels mainly silky, does *not* take a polish and		
(a) the silky feel just predominates	——	**Silty loam**
(b) the silky feel is very pronounced	——	**Silt**

Note

A Sands, loamy sands and sandy loams can be further differentiated
according to grain size.

Coarse sand	– Grit
Sand	– Seashore sand – like sugar grains
Fine sand	– Dune sand – like salt grains
Very fine sand	– Grains just visible

B The presence of much organic matter modifies the cohesive
properties of soil, increasing cohesion in coarse-textured soils and
reducing cohesion in some fine-textured soils. Organic matter also
gives a soil a silky feel.

C Appreciable amounts of calcium carbonate, while not affecting the
sand and silt fractions of soil, do affect the clay fraction tending to
make soils with a significant clay content less sticky than
otherwise.

Soil properties influenced by texture

Texture influences a number of chemical, physical and biological
properties of soil. These include drainage, soil water relationship,
and structure development.

- *Drainage properties*. The nature of a drainage problem and its
  effective solution will depend on soil conditions. Soils of the
  texture groups of sands, sandy loams and loams will have
  macropores while clay loams and clays will have very few of
  the larger pores to permit gravitational water movements
  unless the profile is fractured.
- *Available water capacity*. Soils with the greatest reserves of
  available water and therefore least affected by drought are
  those with medium texture. Such soils have a high proportion

of particles of fine sand, very fine sand or silt grades and the greatest volume of mesopores in the soil mass.

- *Soil structure*. Texture has a considerable influence on soil structure.

Soil structure

The term **soil structure** refers to the fabric of the soil which results from the tendency of individual mineral particles and other soil components to bind together to form **aggregates** or discrete groups of particles. On a larger scale the mass of soil may be divided by a network of recognisable planes of weakness or actual voids into a series of separate units called **peds**. Naturally formed peds are more stable than **clods** formed by soil cultivation. Ped faces are usually recognisable by having a sheen or distinctive colour and they are often coated with materials deposited by groundwater flowing through the voids. Root systems favour the routes between peds and are often crowded on ped faces. The structures occurring in a soil are influenced by the texture, the nature of clay minerals present, the climate and by standards of soil management.

Soil texture has a considerable effect on soil structure because the different soil fractions have different chemical and physical properties. Sands and silts have little natural cohesion and where they constitute the bulk of a soil the profile can appear as a homogeneous mass of single grains. Any structures present are easily crumbled. On the other hand the highly cohesive clays can form a profile which is an unstructured coherent mass, but where structures do form they can be well defined and stable.

The influence of climatic factors on soil structure is of considerable importance for soil drainage and fertility:

- *Mechanical effects*. In heavy rain the raindrops bombard the soil surface, breaking down peds, clods and aggregates often forming an impervious, structureless surface layer.
- *Freeze/thaw effects*. Repeated expansion of soil water as it freezes in cold weather, followed by thawing and the entry of more water can break up the soil mass to the limit of freezing. This process is very beneficial in clay soils by producing a mass of crumbs called a **tilth**.
- *Wetting/drying effects*. Clay minerals, particularly those of the smectite group, tend to swell when wetted and contract when

they dry out. Starting with a fully saturated and structureless, clay-rich soil, as the drying season progresses the soil profile dries from the ground surface downwards and shrinks. Loss of volume causes tension fractures in a roughly hexagonal pattern.

As drying advances into the profile the fractures penetrate more deeply, eventually forming a series of vertical, hexagonal columns in the subsoil. A return of wet weather causes the soil to swell once more and the gaps close up but the discontinuity remains to delimit the columns. In smectite clays the fractures may so improve soil drainage that excess soil water moves away before re-wetting is complete and the voids become a permanent feature of the profile. Fractures may reach about a metre depth in areas with the greatest soil moisture deficit while in cool wet areas they hardly develop at all.

Standards of soil management affect soil structures. Sandy soils are fairly tolerant of badly timed field operations but structures can be created and fertility improved if the soil organic content is kept as high as possible. On the other hand, clay soils are easily damaged by field traffic. Dry clays are very hard, sustaining traffic easily, and fracture into clods if disturbed. With increasing water content, however, they become soft and easily moulded having reached their **plastic limit**. Any field traffic in this soil condition squeezes aggregates together to form a strucureless, impermeable mass that can dry out to be very hard and inhospitable for crop roots. Further increases of water content to a wet clay causes it to reach its **liquid limit** at which point it changes to a slurry. Good management helps to promote good structures while an increase of organic content encourages earthworm activity and the formation of soil aggregates.

Description of soil structures

Soil peds are found in a variety of forms and, to a considerable extent, their nature influences groundflow movements through fine textured soils. A field description of soil structure should include the following terms:

(a) Type – the shape and arrangement of peds.
(b) Class – an accurate assessment of ped size.
(c) Grade – the distinctiveness and durability of peds.

Ped type   A number of distinctive ped shapes are found, as illustrated in Figure 4.2. **Prisms** have the vertical dimension greater than horizontal dimensions. Peds are termed **prismatic** if the tops are angular and **columnar** if they are rounded. **Blocks** have all dimensions approximately equal. If the blocks have sharp edges and corners they are described as **angular blocky**, and this type may be the result of cross fractures occurring in prisms. If the blocks are more rounded the peds are **subangular blocky** and these may be the result of further breakdown of blocky peds. **Spheroids** are small, roughly spherical peds which are found in the topsoil as a result of biological activity, frost action or cultivations. They greatly increase the volume of mesopores in fine textured soils. Such peds are described as **granular** and a particularly porous form is referred to as **crumb**. In the spheroid group the terms aggregate and ped merge to form a gradational series of soil structures. **Plates** have horizontal dimensions greater than the vertical dimension. **Platy peds** are commonly associated with compacted horizons where they form in response to downwards pressure. They form a considerable barrier to the downwards flow of groundwater. If the plates are very thin they may be described as laminar.

Figure 4.2   Ped shapes

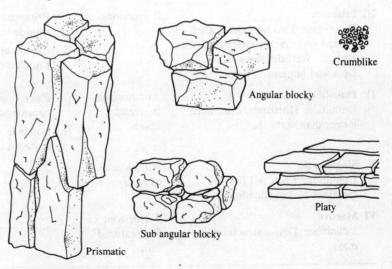

Crumblike

Angular blocky

Platy

Sub angular blocky

Prismatic

Ped class   The **class** of soil structure designates the sizes of peds. Five classes are recognised; very fine, fine, medium, coarse and very coarse. The precise size range of each class depends on the ped type for there is no agreement on actual sizes between different groups of workers. The modified United States system is shown in Table 4.3.

Table 4.3   Soil structure classification

Structure type	Where found	Examples	Ped size (mm) (smaller dimension)
I Spheroidal Definition: Having curved surfaces which have slight or no accommodation to the faces of surrounding aggregates	A Horizon i.e. top soil	(a) **Granular** – relatively non-porous aggregates (b) **Crumb** – porous aggregates	Fine 2 Medium 2–5 Coarse 5–10 Very coarse 10
II Block-like Definition: All axes of approximately equal length, plane or curved surfaces that are casts or moulds formed by the faces of surrounding aggregates	B Horizon i.e. Subsoil	(a) **Blocky** – faces flattened, most vertices angular (b) **Subangular blocky** – mixed rounded and flattened faces with many rounded vertices	Fine 10 Medium 10–20 Coarse 20–50 Very coarse 50
III Prism-like Definition: Two horizontal axes limited and much shorter than the vertical. Well defined vertical faces and angular vertices	B Horizons	(a) **Prismatic** – without rounded caps (b) **Columnar** – with rounded caps	Fine 20 Medium 20–50 Coarse 50–100 Very coarse 100
IV Plate-like Definition: Horizontal axes much longer than vertical	Any horizon – frequently absent	(a) **Platy** – units of considerable thickness (b) **Laminar** – units quite thin	Fine 2 Medium 2–5 Coarse 5–10 Very coarse 10
V Single grain Definition: Each soil particle functions as an individual	All horizons of sandy soil		
VI Massive Definition: Dense structureless mass	B Horizons of fine textured soils		

Ped grade

The **grade** of soil structure or **structural development** refers to the degree of ped or aggregate separation and expresses the difference between cohesion within structures and adhesion between structures. In the field, the durability of structures is assessed in the following terms.

(a) Fine-textured soils

- *None* – there are no ped faces visible, digging is difficult.
- *Weak* – a spadeful of soil (cube) parts with difficulty, only a few ped faces are seen.
- *Moderate* – the cube parts more easily but peds break and the broken surfaces equal formed ped faces.
- *Strong* – peds easily separated by hand when the cube is disturbed.
- *Very strong* – peds separate under their own weight as the cube is placed on the ground.

(b) Coarse-textured soils

- *None* – no ped faces visible but digging need not be difficult.
- *Weak* – the cube shatters when tipped from the spade and there are few definite ped faces.
- *Moderate* – the cube shatters when tipped from the spade but a few peds faces are distinctly visible.
- *Strong* – the peds separate clearly when the cube is tipped onto the ground.
- *Very strong* – the peds separate under their own weight during digging.

This emphasises the point that structures are influenced by textures and there is a strong relationship between the two properties, as shown in Table 4.4.

Profiles without peds

Many profiles and soil horizons show no signs of having structural fractures and two very different types are found.

- *Single grain* profiles in which each soil particle is independent of all others as is the case in many sandy soils, particularly the coarser type.
- *Massive* profiles in which the soil mass has no definite lines of weakness. This is typical of clay soils in wet climatic regions.

Table 4.4   *The influence of texture on structure*

Texture	Topsoil structure	Subsoil structure
Sand	Single grain	Single grain
Loamy sand	Single grain or very weak crumb	Single grain or very weak blocky
Coarse sandy loam	Single grain or weak crumb	Very weak or weak, blocky
Sandy loam	Single grain or weak crumb	Weak, blocky
Fine sandy loam	Very weak or weak, crumb or subangular blocky	Very weak, blocky or platy
Loam	Moderate, crumb or subangular blocky	Moderate, medium to coarse blocky
Silty loam	Weak, granular or fine subangular blocky	Weak, medium to coarse blocky
Silt	Weak or moderate, crumb fine subangular blocky	Moderate, medium to coarse blocky
Clay loam	Moderate or strong, crumb or fine subangular blocky	Moderate or strong, blocky or prismatic
Silty clay loam	Weak or moderate, crumb or subangular blocky	Weak or moderate, blocky or prismatic
Sandy clay loam	Weak or moderate, crumb or subangular blocky	Weak or moderate, blocky or prismatic
Clay	Strong, crumb or subangular blocky	Strong or very strong, blocky or prismatic

Soil pore spaces   The importance of pore-size distribution in terms of soil drainage and fertility has already been discussed. Equally important are the origins and stability of voids in the profile. Pore spaces are the result of:

- Spaces between individual mineral particles as determined by texture.
- Spaces between soil aggregates and peds resulting from soil structures.
- Spaces resulting from biological activity such as channels formed by plant roots and the burrowings of earthworms, moles and other species living in the soil.

Textural porosity is a relatively fixed property, determined by particle size, sands having mainly macropores and clays having mainly micropores. The macropores found in fine-textured profiles can be due only to structural voids and to cavities resulting from biological activity. For this reason, the pore-size distribution in fine-textured soil is variable.

Soil induration

Soil porosity may be reduced because of natural **induration** which involves compression of soil particles and sometimes also the filling of voids by other materials carried in by groundflow. Two quite separate processes are responsible. One results from the conditions of drift deposition. Glacial till and solifluxion layers resulting from glaciation were deposited in a slurried condition and many have dried out to become wholly **indurated profiles** which are hard and impermeable throughout the drift layer beneath a shallow topsoil rendered permeable by biological activity. The other process is the result of **leaching**. Constant downwards flow of gravitational water can move the more mobile soil constituents out of the A horizon and in some soils leached material is deposited in a fairly restricted layer within the B horizon. Such confined zones of limited permeability in a profile are called **indurated horizons** or **natural pans**. Severe leaching is characteristic of the podzol group of soils (see Soil Classification, p. 54) and the process is sometimes called **podzolisation**. Compounds of iron can be leached out of the surface layers and accumulate in a narrow horizon a few centimetres thick to form a hard, red, impermeable **iron pan** somewhere between 35 and 70 cm depth in the profile depending on rainfall and soil permeability. In a similar way, organic decay products, clay and other colloidal materials can form a **clay pan** which is seen as a pale grey to black horizon and is equally impermeable.

Soil compaction

Profile **compaction** is the result of farming activities and has a similar effect to natural induration. Several types of induced compaction are recognised and are fairly widespread in all except coarse-textured soils. Clay and clay loam soils are especially susceptible to this kind of damage since they dry slowly and are often cultivated while in a plastic condition. **Plough pans** and **poaching** have been discussed already. **Soil capping** is the result of the destruction of soil aggregates on the surface by rain drops and is a serious problem because it prevents infiltration and may result in surface erosion. Soil capping is most likely to occur in profiles consisting mainly of silt or very fine sand but it can occur in any soil which is low in organic matter. Bare soils are most often affected and capping hinders soil aeration, seed germination and crop emergence in addition to reducing infiltration.

Soil permeability

The dry weight of a unit volume of soil in its field condition is referred to as the **bulk density**. Since soil particles are relatively uniform in relative density, a soil bulk density value is a function of the packing arrangement of particles. Friable, well-aerated soils have a low bulk density while compacted soils have a high bulk density. The rate at which groundwater çan move through a profile is described as the **hydraulic conductivity** or **K value**. This property is rarely uniform throughout even a small area of land because of soil variations and each horizon may have its own distinctive value. Vertical hydraulic conductivity may be very different from the horizontal values. This soil property is a function of the nature and arrangement of soil particles and as a result there is a close relationship between pore-size distribution, bulk density and hydraulic conductivity in a soil. Very often the hydraulic conductivity is a function of texture, ranging from high values in coarse sands to low values in clay. For a given texture and bulk density, hydraulic conductivity is a relatively constant property. It can be greatly increased by soil structural voids and greatly reduced by induration or compaction.

Soil colour

One of the most obvious characteristics of a soil profile is the variation of colours in the different horizons and this is important because it reflects drainage status, organic content, mineralogy and soil morphology. The main factors influencing soil colours are as follows:-

Iron compounds, though not often present in great quantity, have a marked influence on colour because they appear as coatings or stains on particle surfaces. The most common compounds of iron are ferric oxide which is red; hydrated ferric oxide which is yellowing brown; and ferrous oxide which is bluish grey. The particular oxide of iron present in a soil depends on the effects of oxygen in the soil mass and, as a result, the colour of a profile may be indicative of its long-term natural drainage status. In freely drained soils oxides of iron are fully oxidised to the ferric state and the colour tends to be in the range of brown, yellow or red. Poorly drained soils lack oxygen and iron occurs in a ferrous state which confers medium to dark grey colours often tinted bluish or greenish. Horizons which are saturated most of the time, but occasionally are not saturated, appear mottled with flecks of bright colour (brown, red, orange) against a dull background. The bright colours are often near to where roots have penetrated or where other voids have allowed the entry of air.

Humus and organic breakdown products, like iron compounds, tend to coat soil particles and impart brown or black staining. Only about 5 per cent of organic matter in a profile can give the soil a brown or black appearance.

Most other minerals originally are white or light grey in colour and become stained by iron compounds or organic material. A few, however, have distinctive colouring which is seen in profiles and may mask the effects of pedological processes. Grey parent materials (such as marine alluvium) may conceal aerobic conditions in a freely drained soil whereas red parent material (such as Old Red Sandstone) may conceal anaerobic conditions in a poorly drained soil. Hence, soil colour may be indicative of soil drainage status but the evidence should be considered with other observations.

Iron ochre in soils

The conversion of ferrous iron compounds into a ferric state when air is introduced into a profile is of interest to a land drainer for more immediate reasons than change of profile colour. Ferrous oxides are more soluble in groundwater than ferric oxides and when the profile becomes aerobic further oxidation causes ferric oxide to precipitate out of solution. This reaction is initiated and speeded up by some species of soil bacteria. In most soils the process is hardly noticed but in profiles that are unusually rich in iron, the ferric oxide or **iron ochre** appears in such quantities that drainage installations may be blocked or the profile nearby rendered impermeable. Soils derived from basin peats or from rock waste rich in iron are particularly difficult to drain for this reason. Most sediments of Carboniferous age come into this category.

Soil organic matter

Biological activity produces a wide variety of discarded tissues and waste products which tend to accumulate on or within the topsoil. Further biological activities (particularly by earthworms, fungi and micro-organisms) then convert the organic debris into a well-decomposed brown or black, jelly-like substance called **humus** and thoroughly incorporate it into the soil mass. This improves soil fertility by promoting a better pore-size distribution and by releasing plant nutrients. Organic matter constitutes between 1 and 10 per cent (of oven-dry weight) of the topsoil weight (20 cm depth) of most soils. Soils with less than about 5 per cent of organic matter may have fertility limitations while those with more than 20 per cent are described as organic soils and have distinctive properties compared with mineral soils.

Stones in soil profiles

The stoniness of a soil influences its value for agriculture. Large stones, greater than about 25 cm, create serious problems with cultivating implements and increase production costs. If present in appreciable quantities (3 per cent or more of the soil mass) they hamper farming operations considerably. This includes land drainage work which may be rendered more difficult in terms of accuracy of installation or more expensive because of implement wear and the subsequent cost of stone removal from spoil. A large stone content may render land improvement less worthwile.

Soil classification

There are several comprehensive soil classification systems based on comparisons of the soil horizons. These soil variations are the result of the close relationship that exists between the climate, natural vegetation and soil development which is reflected in the characteristics of the A and B horizons. Of particular importance to the profile development is the balance between rainfall and potential evapotranspiration. Where the potential evapotranspiration exceeds the rainfall any dissolved substances in the groundwater will tend to accumulate in the upper soil horizons and may precipitate out as solid concretions. On the other hand, where rainfall exceeds evapotranspiration any substances in the topsoil that are soluble or otherwise mobile may be leached out. Very many types of soils are listed on a world classification but in the middle latitudes four broad groups dominate and form a gradational series. The various names mentioned in each group come from different soil classification systems but are more or less equivalent terms. Starting in the dry continental interiors and moving towards the wet coastal margins the following broad types of soil are met in turn.

1 In the arid areas where the solum is dry for most of the year desert soils are found, collectively called **aridosols**, **sierozem**, **solonetz**, and are the result of the potential evapotranspiration far exceeding the rainfall. Where a supply of groundwater or regular irrigation allows an increase of the actual evapotranspiration, salts may accumulate in the surface horizons to form **salty soils** or **solonchak**.

2 The **mollisol**, **chernozem**, **brunizem** or **prairie** soils are a soft friable group that are found in the prairie grasslands where there is a general balance between rainfall and potential evapotranspiration. Organic materials from the grassland community accumulate in the A horizon to give it a dark colour and nodules of calcium carbonate may develop in the B horizon.

3 The **brown forest soils** or **alfisols** occur where rainfall exceeds evapotranspiration by a moderate amount. Some leaching occurs but the natural vegetation of deciduous trees takes up plant nutrients from the soil in large amounts to keep them in circulation. The soil is a rich brown colour with indistinct horizons because of earthworm activity.

4 The **grey forest soils, podzols** or **spodosols** occur in the wetter cooler areas where rainfall considerably exceeds evapotranspiration and leaching is severe. Plant nutrients are lost and the climax vegetation on the poor acid soils is coniferous forest. Earthworms are absent from the inhospitable soil and a dark layer of plant litter accumulates on the surface. Beneath this layer is the characteristic grey horizon from which all staining materials have been leached. The leached substances are found often as pans or distinctive horizons within the B horizon.

Figure 4.3 Soil profiles

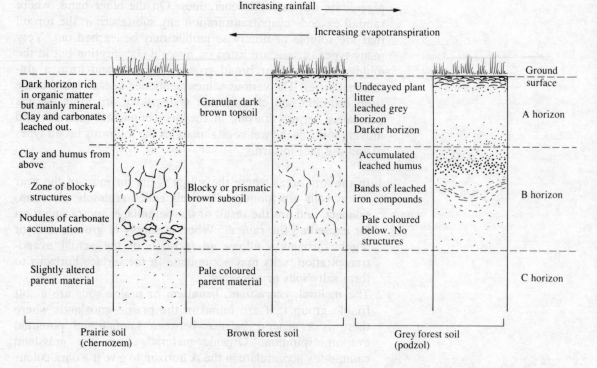

Increasing rainfall →

← Increasing evapotranspiration

Dark horizon rich in organic matter but mainly mineral. Clay and carbonates leached out.

Clay and humus from above

Zone of blocky structures

Nodules of carbonate accumulation

Slightly altered parent material

Granular dark brown topsoil

Blocky or prismatic brown subsoil

Pale coloured parent material

Undecayed plant litter
leached grey horizon
Darker horizon

Accumulated leached humus

Bands of leached iron compounds

Pale coloured below. No structures

Ground surface

A horizon

B horizon

C horizon

Prairie soil (chernozem)

Brown forest soil

Grey forest soil (podzol)

The various types of profiles are shown in Fig. 4.3. There are large areas with intermediate types and any cultivation mixes up the characteristic features of the A horizon.

Wetland soils In the geographical soil classification described above only those found on the freely drained sites are considered. When the gradational series is carried into regions of even greater rainfall and less evapotranspiration the profiles become increasingly wet. The same is true elsewhere if the soil is situated in a hollow with impeded drainage. All soils of wetland sites are described as **gley soils**. Gley soils have developed in anaerobic conditions and display the characteristic bluish grey, olive grey or dull grey tints and are often sticky or compacted. Individual horizons with such characteristics are described as **gley horizons**. This term should be taken to include mottled horizons of alternating aerobic and anaerobic conditions. With time when the soil is saturated to the

Figure 4.4   Soil profile changes

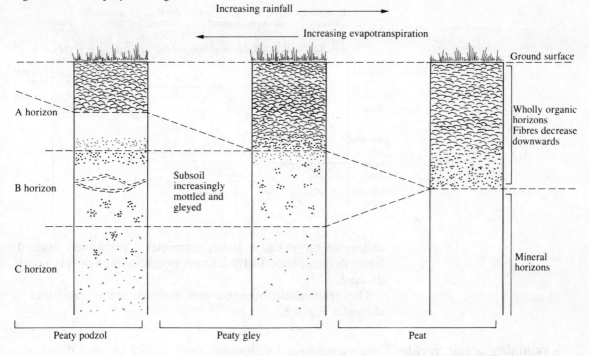

Increasing rainfall ⟶

⟵ Increasing evapotranspiration

Ground surface

A horizon

B horizon

C horizon

Subsoil increasingly mottled and gleyed

Wholly organic horizons
Fibres decrease downwards

Mineral horizons

Peaty podzol   Peaty gley   Peat

surface, the amount of organic waste collected at the surface increases and podzols grade into **peaty podzols**, **peaty gleys** and finally into **peat** of increasing depth. This series of profile changes is shown in Fig. 4.4.

Soil surveys   Various national soil survey projects have made considerable progress in mapping soils. Generally they use the broad geographical classifications described above but take account also of the soil parent material. The Soil Survey of Scotland groups all soils derived from the same parent material as a **soil association**. Within each association the effects of the soil drainage status produce distinctive profiles and each is described as an individual **soil series**. The soil series is used as the basis of soil mapping and each is comprehensively described. Soil maps, where available, are valuable aids to the drainage designer and should be used in conjunction with careful examination of the soil profiles in the field. As already discussed, a soil series is assigned

Figure 4.5   Classifying profile drainage status

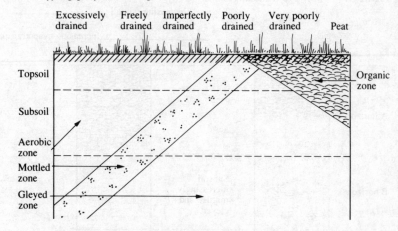

to one of the drainage status categories – excessively drained, freely drained, imperfectly drained, poorly drained or very poorly drained.

The relationship between soil drainage status categories is shown in Fig. 4.5.

Examining a soil profile

Soil examination for drainage work should be carried out in a systematic manner to determine soil properties, drainage status and the best drainage solution. The following procedure is suggested.

1 Survey the whole area noting the topography of the site and surrounding area. Within the area establish any pattern of soil variation by means of a soil auger or spade and where more than one distinctive soil type is present, mark the boundaries on a field sketch map.

2 Open a small number of pits, at least 2 m deep, across the area using a mechanical excavator so that a full range of profile types is exposed. Selecting a typical profile for each main soil type present, remove the smeared margins of the pit with a spade or trowel to reveal the undisturbed horizons.

3 Note the general depth of the solum and the stoniness of the profile. If the soil layers are generally permeable note the depth to any bedrock or other less permeable layer or formation below the solum. If the soil layers are not generally permeable

note the depth of any more permeable formation below the solum (see chapter 9).

4 Carefully assess the textural class of topsoil and subsoil. The most important zone is the B horizon above the likely level of land drains, usually about 60 cm below the ground surface.

5 Note the type, class, grade and depth of penetration of any soil structures in the subsoil.

6 Examine the profile for indications of induration or compaction. These layers are either hard when first observed or become hard as the walls of the pit dry out. They are easily identified by probing with a trowel or knife. Note the depth at which induration or compaction occurs and the extent of the impeded zone within the profile. Bear in mind that macropores are visible and easily identified when present in a soil mass.

7 Consider the drainage status of the profile. If the inspection pits in the area intercept a groundwater table, water will fill the holes to the level of the groundwater table. If a perched water table is present, water will seep into the test pits from the saturated zone above an impeding horizon. In a dry season the presence of gleyed horizons and other evidence of the wet season drainage status may be noted. A well developed system of roots penetrating deeply into the B horizon indicates that, at the time of growth, the drainage status was satisfactory. Root systems confined to the A horizon, growing horizontally and forming a mat above some barrier, indicate poor permeability or poor drainage status below that level. Organic matter like ploughed down turf, crop residues or manure that has not decomposed within the profile and gives off the unpleasant smell of hydrogen sulphide indicates anaerobic conditions which may be caused by lack of macropores or waterlogging. Similarly, the movements of earthworms are restricted by compacted or saturated horizons and all exposed horizons should be examined for their presence.

8 By observing all of the holes note the evidence of any groundflow in the site. Groundwater may flow into the pit from a deep permeable layer intercepted by the excavation. Note the level to which the inflowing groundwater rises in the pit after a few hours. Evidence of groundwater flowing across the site may be obtained by noting the level of groundwater in all of the pits from a common datum level and noting any gradient in the groundwater table.

Assessing profile hydraulic conductivity

In most cases examination of a soil profile will permit a reasonably accurate estimation of its present and potential values of hydraulic conductivity. For assessment purposes it is necessary to identify the soil texture class fairly accurately and to take account of the size of sand grains present – that is, to identify the sand fraction as being coarse, medium, fine or very fine. The most likely correlations between non-fractured (that is, massive or single grained), non-compacted texture classes and K values are shown in Fig. 4.6.

The K value can be greater than that suggested by texture classification if fractures are present to a significant depth in the B horizon. Generally the K value increases as ped size decreases (ped class), as ped strength increases (ped grade) and as depth

Figure 4.6   Assessing horizon hydraulic conductivity values

Scale of profile hydraulic conductivity in m/day

Likely hydraulic conductivity values to be found in non-indurated, non-compacted, unfractured profiles based on texture classification:

Hydraulic conductivity can be reduced by a significant amount, often to zero by compaction or induration:

Hydraulic conductivity can be significantly increased by structural voids, particularly in fine textured profiles:

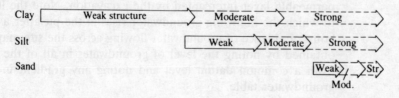

of penetration increases. Structures have greatest effect in fine textured soils and have a diminishing influence as the dominant fraction particle size increases such that they have little effect on coarse sands. A general indication of possible adjustments for structure are shown in Fig. 4.6. It is advisable to be cautious when making allowance for structures. As a rule, make full allowance for the effects of fractures only where they are a permanent feature, are strongly developed and penetrate deeply into the B horizon. Where weaker or shallower fractures are found it is safer, usually, to make a lesser adjustment and where fractures are observed to close up during a wet season it is better to make no adjustment at all.

Hydraulic conductivity values can be much reduced by compaction or induration. Where only a narrow zone in the profile is affected, for example by a pan which is easily disrupted, it can be ignored when assessing hydraulic conductivity. If the whole profile is indurated it is necessary to assess the K value before disruption (usually zero) and to assess the potential K value after disturbance as indicted by texture class.

Measuring soil properties

The likely behaviour of soil water is the most important factor to be assessed in any soil examination for drainage design. A variety of experimental methods, equipment and techniques have been developed for measurement of the soil properties involved. They may be divided into those that measure:

- *Water levels* – various devices that seek to indicate the actual or potential level of free (gravitational) water in the soil profile.
- *Soil water tension* – apparatus to measure the degree of the soil water deficit above the groundwater table.
- *Bulk density and pore spaces.*
- *The infiltration rate.*
- *The hydraulic conductivity.*

Except in special cases, detailed experimental observations are more suitable for research and development programmes than for field drainage schemes. Because of the costs and time-consuming nature of the work and the necessary planning and interpretation required with these methods it is unlikely that they will find widespread application on a routine basis. Results from detailed programmes, allied with climate information and other aspects of field drainage, may be adopted to simplify or organise field assessments of drainage. As it is, the observations and interpretations of soil properties available to the man in the field are a more practical option and with care this method need not be less accurate in terms of drainage improvement. In this respect, the hydraulic conductivity of a soil is very difficult to measure because of soil variations through and along the profile, especially where fracturing has occurred. Assessed values are often more accurate than figures obtained by apparently precise measurement.

Drainage problems associated with profile characteristics

Soils with a coarse texture (those composed of sand, loamy sand, sandy loam and the equivalent classes of coarse sand) have a large proportion of macropores and are highly permeable. Profile drainage characteristics depend on site position in the landscape and, therefore, are a function of topography. Those on elevated sites are freely or excessively drained while those in hollows may be very poorly drained. This depends on the depth below the surface of the groundwater table and permeable profiles with a persistently high groundwater table are called **groundwater gleys**.

Soils with a fine texture (those composed of sandy clay loam, clay loams and clays) have a large proportion of micropores when in a massive condition and are either slowly permeable or impermeable. Profiles lacking interconnecting macropores cannot have a groundwater table. Those profiles that have structural fractures will be massive beneath the depth of penetration of fractures in the subsoil. Poor drainage characteristics are very common and occur as a perched watertable in the permeable topsoil and in the upper horizons of the subsoil to the limit of penetration of structure fractures. Such a profile is described as a **surface water gley**.

Soils with a medium texture (those composed of fine sand, very fine sand and their loamy equivalents together with loams and silt) have a mixture of pore sizes or have mainly mesopores and are moderately permeable. They have intermediate drainage characteristics and the profile may be either a surface water gley or a groundwater gley if poorly drained.

Soils of any texture which have an impermeable horizon in the profile can have the symptoms of a surface water gley.

Soil management   The water content of soils, the level of the groundwater table and the persistence of waterlogging are all features which can fluctuate readily, depending on the climate, crop growth and standards of land management. Soil management practices should be designed to give plants the optimum growing environment by providing a soil which contains available water and air spaces. At the same time the soil should be easily penetrated by the growing roots yet firm enough to carry implements and animals without being damaged. Management practices must seek to match the soil water content to the crop requirements. In many areas of the world this may involve considerable effort and capital expenditure. Broadly speaking, **irrigation** by spraying or by deliberate flooding is needed to supplement the natural rainfall when a soil water deficit is a major limiting factor for crop growth, whilst in wet regions **land drainage** is needed to remove excess, superfluous or gravitational water to allow the entry of air into the soil voids. Soil drainage may be improved by lowering the groundwater table, intercepting any inputs of lateral flow or by increasing the movement of excess water in the soil by improving the permeability of the soil mass. Many poorly drained soils can be improved by **soil treatment** either alone or in conjunction with a new drainage system. In fine-grained profiles the success of the drainage work depends on the creation of fairly stable macropores to sufficient depth in the profile to remove excess water from the rooting zone. It is important to bear in mind always that soils with an appreciable clay content are too wet at field capacity to withstand traffic. Drainage improvement, by itself, cannot reduce the soil water content below field capacity. Further drying out by evaporation or transpiration is needed before a clay soil can be safely cultivated or grazed.

Chapter 5 **Organic soils**

Organic matter in soils

All soils contain some organic matter as a result of biological activity. The amount found in any profile depends on a number of factors but essentially it is controlled by the difference between the rate of production of new plant tissues (plant growth) and the rate of decay of discarded tissues (annual die-back and fallen leaves). If the growth rate exceeds the decay rate then profile organic content increases and, conversely, any factor which accelerates decomposition or reduces the supply of plant materials will lower the organic matter content. Most natural profiles in temperate latitudes contain between 5 per cent and 10 per cent organic matter in the A horizon. In farmland the amount present can be increased or maintained by including leys in the rotation and by adding animal litter, crop by-products or other organic wastes to the soil. It is reduced by continuous arable cropping and by burning crop residues. Infertility increases as organic content drops below 5 per cent in the topsoil. In environments where plant growth rate naturally exceeds decay rate a layer of plant debris accumulates on the ground surface and this layer varies in thickness from the merest trace to an organic layer many metres deep.

Classifying organic soil

There is no clear distinction between mineral soil and organic soil, as illustrated in Figs 4.4 and 4.5, and some conventional boundaries must be adopted. Organic layers less than about 10 cm deep are characteristic of podzols and such profiles are clearly in the mineral category. With greater depth of organic material, associated with a poorer drainage status, the terms peaty podzol or peaty gley must be used as appropriate. When such profiles are cultivated the layers are mixed to form an intermediate type of Ap horizon. Organic layers of 30 cm or more are described as **peat**. Some natural environments like river flood plains can cause plant debris and rock waste to accumulate at the same time forming a natural mixed profile. Such horizons with between 10 per cent and 20 per cent organic matter are described as **humose** while those between 20 per cent and 45 per cent are described as **peaty**. These terms can be used in a textural classification as required to describe, for example, humose silt or peaty sandy loam. Horizons with more than 45 per cent organic matter are called **peat**. Thus, use of the term 'peat' should imply a layer on the surface of at least 45 per cent organic matter not less than 30 cm deep. However, the term is often used, wrongly, to describe all types of organic layers. In respect of land drainage work a profile with 30 cm of peat is a very different proposition from a peat formation 30 m deep but there are no descriptive terms or mapping systems to indicate peat depth. The term **histosol** is used to describe organic soils in one soil classification system. The names used are neither precise nor universally adopted.

Peat accumulation   In most environments tissue decay keeps pace with the arrival of more organic material as part of a cycle of biological activity. Changes of climate with the seasons affect both sides of the cycle equally and increased growth rate is matched by increased activity in the soil. But where soil saturation or extreme soil acidity occurs it is likely that plant growth, however restricted, will exceed the tissue decay rate to some extent and peat will accumulate. An anaerobic and/or acid environment suppresses the tissue decay organisms. Wherever the land surface is permanently wet it is likely that peat will accumulate. This may be caused by climate, by topography or by escaping groundwater.

The influence of climate   The effect of climate on peat accumulation can be illustrated by considering the range of soil types of the maritime temperature regions in sequence. The sequence discussed may relate to a traverse of soil types travelling from better to poorer climatic conditions or, equally, to a single site experiencing a deteriorating climate over a long period of time.

In Prairie and Brown Forest soils any plant debris is quickly incorporated into the profile and no very distinctive organic layer develops. An intimate mixture of mineral and organic matter called **mull humus** is found typically under prairie grassland, good cultivated pastures and deciduous forests. The characteristic brown colour is most marked near the ground surface and diminishes with depth. In a colder wetter climate podzolic soils are found. The leached acid soil inhibits the most efficient breakdown organisms and pine needles are less suitable as a food source for earthworms and micro-organisms. Plant litter is only slowly broken down and incorporated by less efficient organisms, resulting in a distinct organic horizon of **mor humus**. Mor humus is a dark brown or black surface layer up to 10 cm thick and clearly distinguished from the horizons below. An even colder and wetter environment will cause poor soil drainage. Podzols grade into gley podzols and then to peaty podzols, finally into peat as increasing acidity and lack of soil air effectively destroys all organisms capable of rapid breakdown of organic wastes. Plant growth, however slow it becomes, always exceeds tissue decay and year by year, over the centuries, the layer of peat becomes thicker. Peat accumulations of this type are called **high moor**, **blanket bog** or **blanket mire** and the less specific term **hill peat** usually implies that it is a peat accumulation caused by climatic factors. The term 'blanket' is particularly apt as it describes the manner in which the peat covers every undulation of the surface like a blanket.

The vegetation of blanket mires

There are two basic types of blanket mire vegetation. Lichens, mosses, low growing herbs and woody plants survive in the very cold tundra where the water comes only during the summer melt. In the more familiar temperate blanket mires, constantly wetted by rainfall, the sphagnum/cotton sedge community is widespread. Heather, in stunted form, is confined to the drier, more elevated hummocks. On active mires, that is where the peat is still accumulating, sphagnum mosses are dominant forming **sphagnum peat** as they grow year by year. A change to drier weather alters the balance, and where the surface can dry out for short periods the sphagnum is inhibited and replaced by heather as the dominant species. Many areas of heather moor in the mountains and uplands of the maritime regions are former blanket mire environments indicating a long-term improvement in climate. In this environment the peat is not accumulating.

The influence of topography

A second factor causing peat accumulation is restricted surface drainage in depressed parts of the landscape. Land surfaces created by water erosion tend to have very few areas of restricted drainage apart from a narrow zone near the channel in valley bottoms. The majority of peat-filled hollows are associated with glaciated landscapes. The melting ice sheets left an undulating surface with many water-filled hollows, of which only the deeper ones can still be seen by the casual observer. The shallower examples are now completely filled with sediments and peat. This type of peat accumulation occurs wherever standing water causes permanent saturation of the surface layers of the rock-waste. The peat formed by restricted surface drainage is broadly described as **low moor**, **basin peat** or **basin mire**.

In maritime regions all hollows are filled with water and overflow throughout the year to form a connecting system of streams. Water flowing into a water-containing hollow carries in rock waste which is deposited in the standing water and collects on the bottom as a sedimentary formation. Water overflowing from the hollow slowly erodes the retaining rim (especially if it is composed of glacial drift) and lowers the water surface. At the same time aquatic and swamp plants colonise the site and further reduce the depth of water by trapping mud and adding plant debris to the deposits. Eventually the whole area of ponded water becomes a reed swamp. The material collecting at this stage is a mixture of mud and the discarded parts of swamp plants and peat formed in this environment is variously described as **sedge peat** or **phragmites peat**. All such peat accumulations are watered by surface flow so that they are enriched by mineral substances carried in by the water and may be described as **minerotrophic mires**. Depending on the nature of the inflowing water, these mires and the resultant peats may be acid or basic in character. Where the water flows over chalk, limestone or other calcic rock formations before entering the swamp area the mire is basic and is described as a **fen** while the organic material that accumulates is **fen peat**. However, the meaning of the term 'fen' is sometimes extended to include all minerotrophic peat and, therefore, basin peat as opposed to hill peat. When the accumulating peat and water-borne sediments build up to the general level of the outflow channel the standing water is no longer evident, apart from a narrow tract of water sometimes seen to mark the line of stream flow. The bulk of the site is now a mire or peat-filled basin. If the climate of the area is relatively mild and dry a new plant community can develop.

Continuing channel erosion by the stream can lower the groundwater table sufficiently to allow the establishment of grasses, shrubs and trees suited to the saturated soil and the sedge community becomes restricted to the minor hollows within the site where the groundwater table remains above the ground surface. Purple moor grass, willow and alder are common species in this type of environment. Mires with a cover of shrubs and trees are called **carr** or **fen carr** and any peat formed at this stage is described as **grassy peat**, **woody peat** or **carr peat**. Eventually, continuing channel erosion may so lower the groundwater table that the surface of the peat can dry out bringing peat formation to an end and allowing the site to be colonised by a coniferous tree community.

Raised mires

Where a peat-filled hollow is sited in an area with a cool temperate climate a different sequence of events will occur. Peat holds more surface moisture than mineral soils and where low levels of evapotranspiration ensure continued waterlogging of the surface the necessary anaerobic conditions for further peat accumulation are provided. Once the surface is lifted above the groundwater table the new horizons of peat become dependent on direct rainfall as a source of water and are no longer modified by any inflow of mineral substances. This creates a much more acid and less fertile environment for plant growth and the reed swamp community is replaced by the sphagnum/cotton sedge community of blanket mires. The build up of peat continues and eventually an elevated dome of peat is formed which grows upwards and outwards far beyond the original water-retaining hollow. Peat formations of this type are called **mosses**, **raised mosses**, **raised bogs** or **raised mires** and show many variations depending on the climate and the slope of the land surface. They grow most rapidly in the coolest, wettest regions where they merge with and are indistinguishable from blanket mires. Raised mires and blanket mires are very acid, lack mineral substances and both receive all moisture in the form of direct rainfall. For this reason they are grouped together as habitats producing **ombrotrophic** mires. Raised mires have been known to become mobile in severe storms and flow out suddenly to cover the surrounding countryside with liquid peat. Most of the raised mires situated in less remote regions have long been used as a source of fuel and peat cutting has greatly reduced their extent. The peat obtained is sphagnum peat like that of blanket mires.

Raised mire profiles

Whole sections through a raised mire reveal that the peat contains a number of distinctive horizons, all reflecting the conditions prevailing when the peat was formed. The lowest part of the profile which accumulated in standing water consists of a mixture of mineral sediments and the remains of swamp plants (the minerotrophic horizons). Above this is the purely organic zone formed by the preserved remains of mosses and cotton sedge of the raised mire community constituting the ombrotrophic horizons. Within this upper part of the profile – where it is complete – there is a catalogue of the climatic changes that have occurred in the post-glacial period. Since the end of the most recent glaciation of the temperate zone the climate has continued to be variable. A major cycle of climatic change, each phase lasting many centuries, has brought periods of relatively warm dry climate, called the **Boreal** climatic periods, alternating with colder, wetter **Atlantic** climatic periods. These changes are clearly marked in the raised mires. Peat accumulated and raised the surface level continuously in Atlantic conditions but a Boreal phase brought peat accumulation to an end and allowed the development of an organic soil. As the dry phase continued a succession of vegetation changes resulted in the establishment of a coniferous forest. Eventually the cool, wet weather returned with the next phase of the climatic cycle and the resulting soil saturation killed the trees and re-activated peat growth. The new layers of peat then buried and preserved the remains of the forest and the soil it grew in. Peat cutting has revealed these layers of stumps, fallen trees and associated peat soils separated by layers of fresher peat in all the major mosses where the two layers of 'bog oaks' are a notable feature.

The present environment

Most raised mires away from the very wettest regions are at present inactive, indicating a dry climate period or Boreal phase. Eroding remains of raised mires are found in many lowland areas of the temperate regions where peat formation would not be expected in present climates and many areas of hill peat are eroding and wasting. The drying process has been accelerated often by drainage work but climate is the major factor. Where farming activities prevent the natural return of coniferous forest the unimproved mosses are colonised by heather.

Coastal peatlands

Minerotrophic peat may develop in low lying coastal areas where drainage is restricted by a high sea-level. Peat accumulation goes through all the phases of reedswamp, fen, fen carr vegetation but the process is sometimes interrupted by changes of sea-level. A falling sea-level accelerates drainage and encourages dry land species while a rising sea can drown the peat bog and cover it with marine sands or clays. Falling seas once more allow the erosion of gullies through both mineral and peat layers which may later be filled by further sediments. Meanwhile rivers and streams crossing the area can keep channels clear of peat or may deposit sinuous ridges of sand. Coastal peats, therefore, may be separated horizontally by a layer of marine sediments or they may be interrupted laterally by mineral filled channels and sand ridges.

The influence of springs

The third factor causing peat formation is the surface saturation caused by a persistent spring. The peat that develops in the zone of saturation depends on the mineral content of the spring water but it is usually of the minerotrophic type derived from a wide range of plant species including mosses where conditions are appropriate. In the drier climatic areas the zone of saturation may end fairly abruptly and the peatland community merges with dry land vegetation through a fringe of rushes. A notable feature of such **spring mires** is the commonly occurring steep slope of the surface. Where the spring water emerges in a hollow a floating mat of peat and other vegetation may produce a **quaking mire**. Spring mires can occur anywhere, including the middle of other types of peat bog, and where the spring issues below the peat are either numerous or vigorous they create a particularly unstable surface and an area of peat that is unusually difficult to drain.

The location of peatlands

Blanket and raised mires occur where there is considerable rain-fall throughout the year and temperatures range from cool to very cold. Blanket mires are typical of upland oceanic regions in countries like Norway, Iceland, Ireland and Scotland and where the climate is particularly wet they extend to land at sea-level. Basin mires occur wherever there is sufficient rainfall to prevent surface hollows from drying out. Areas of coastland peat caused by a relatively high sea-level are found in Florida, the Netherlands and the Fenlands of eastern England. Taking all types of peat into account, Finland has most with over 30 per cent of its surface covered. Other major areas are the tundras of Russia and North America. Peat is important also in Iceland, Ireland, Scotland, Norway, Sweden and in the glaciated areas around the Great Lakes in North America.

Soil properties

The properties of organic soils depend on a number of factors relating to the organic material present in addition to the influence of any mineral material it may contain or overlie. These properties depend on the depth of the organic layer, the type of organic material present, the degree of decay and the amount of structural development. Shallow organic horizons on the surface are considered as part of the profile of mineral soils and are described accordingly. Where there is an appreciable layer (over 10 cm) as, for example, in a peaty podzol, several horizons can be described:

- **L horizon** – the top layer of undecomposed plant debris
- **F horizon** – an underlying horizon of partially decomposed plant debris
- **H horizon** – well decomposed organic material overlying the mineral horizons.

Below the organic horizons are the normal A, B and C horizons in undisturbed sites. When the peat is deeper there comes a point where the soil is totally organic and the mineral horizons below are not actually the parent material and should not be considered as B or C horizons. The depth of the organic horizons influences many soil properties and is clearly of fundamental importance in any drainage work.

- In **humose soils**, especially those that contain just enough organic matter for inclusion in the group, the soil properties are determined by the nature of the mineral horizons. With increasing organic content, however, the effects of texture are increasingly masked and structures are modified. In the A horizons sandy soils are aggregated and clay soils are more friable. In the B horizons structures are less well developed than in equivalent truly mineral profiles.

- In **peaty soils** the cultivated horizon is largely organic with properties determined by the nature of the peat. The underlying mineral subsoil will nearly always be a massive gley unless previously improved. In both humose and peaty soils drainage design will be influenced by the nature of the mineral subsoils but with some allowance for the moisture-retaining organic topsoil.

- In **deeper peat** profiles all soil properties are determined by the organic material. The kind of difficulty encountered in land drainage work is strongly influenced by the depth of the peat and much depends on whether drains can be laid on a stable mineral foundation or in the less stable peat.

The composition of peat

The composition of the peat is determined by the environment producing it. Different peat-forming materials have different characteristics and affect the properties of any soil profile that develops. In general terms three broad types of peat can be recognised based on the plant community producing the waste materials.

(a) **Mossy peat** is a very common type found in all blanket mires and raised mires (the ombrotrophic peats). The most widespread mossy type is sphagnum peat although other types are found, e.g. hypnum peat based on another family of moss plants.

(b) **Herbaceous peat** is found in minerotrophic environments and consists of the remains of grasses and swamp plants. This group includes sedge peats and grassy peats and the horizons are often rich in mineral material.

(c) **Woody peat** is found in mires of the carr type and consists of herbaceous peat as described above but with more than 10 per cent of woody material derived from the tree communities.

The type of material present largely determines the peat density and the dry matter content. Dense herbaceous and woody peats can have up to 15 per cent dry matter while mossy peats may have as little as 3 per cent dry matter. This factor influences the water-retaining properties of the peat and the amount of shrinkage likely to occur when water is allowed to drain out. As a general guide, land improvement is more likely to be successful in peats with a high dry matter content.

Humification

Peat consists of the fibres of plant tissues. Usually only the harder parts are preserved for any length of time and the softer tissues are slowly converted into colloidal humus which soaks into the cells of the fibres, helping to preserve them. Depending on conditions the hard tissues like lignin and cellulose may be preserved indefinitely or there may be a very slow decay of the fibres into humus. This process is called **humification**. Cellulose is converted into a humified residue with the release of carbon dioxide, water and methane (marsh gas). Humification is very slow in unimproved mire conditions and only a small amount of peat accumulation is needed to make up the losses. Usually, however, the newest (upper) layers of peat are less humified than the lower layers. The **degree of humification** of peat is expressed as the proportion of fibres present in a sample and fibres are taken to be fragments of plant tissue at least 0.1 mm long. Peat may be classified according to the degree of humification as follows:

- **Fibric peat** has at least 66 per cent of its bulk composed of fibres.
- **Mesic peat** has between 33 and 66 per cent of its bulk composed of fibres.
- **Sapric peat** has less than 33 per cent of its bulk composed of fibres.

For drainage purposes it is useful to classify peat humification in the field and for practical reasons these three types are quite satisfactory although some tests have been devised to give *ten* separate categories. Estimates of the degree of decomposition are made by squeezing a handful of freshly dug wet peat.

- **Fibric** peat releases a clear to turbid liquid and plant structures are easily identified. Only a little peat squeezes between the fingers.
- **Mesic** peat releases a very turbid liquid and up to a half of the

peat oozes between the fingers leaving a mushy, friable residue.

- **Sapric** peat allows at least half of the peat material to squeeze between the fingers as a mush and very few or no clearly recognisable fibres are left in the hand.

Soil texture   Soil texture is a property not strictly relevant to organic horizons. In the case of humose and peaty horizons, the mineral subsoil can be classified as discussed in Chapter 4. The organic topsoil may be classified according to the mineral texture and prefixed by the words 'peaty' and 'humose' taking care to allow for the different 'feel' of the organic constituents when doing so. A system for classifying topsoil texture is shown in Table 5.1.

Table 5.1   Assessment of soil texture (organic soils)

1. The soil is composed mainly of mineral fragments but is dark in colour and stains the fingers, a silky feel can be detected.	Yes No	2 5	
2. The sample is distinctly gritty	Yes No	3	**Humose Sand**
3. The sample is slightly gritty	Yes No	4	**Humose Sandy Loam**
4. The sample is not gritty	Yes		**Humose Loam**
5. The soil is composed of organic and mineral material but is distinctly 'peaty' in appearance and feel	Yes No	6 8	
6. The sample is gritty	Yes No	7	**Peaty Sandy Loam**
7. The sample is not gritty	Yes		**Peaty Loam**
8. The soil is composed mainly of organic material but plant remains are not clearly visible (they are decomposed)*	Yes No	9 11	
9. The sample is gritty	Yes No	10	**Sandy Peat**
10. The sample binds when worked between fingers	Yes No		**Loamy Peat** **Peat**
11. The soil is composed of visible plant remains*	Yes		**Raw Peat**

* Assess the degree of humification

Soil structure

Soil structure is also much influenced by the amount of organic matter present. In profiles with a humose topsoil, structures in the subsoil are as would be expected in a mineral soil of the same texture but perhaps slightly less well developed because surface organic matter inhibits subsoil drying. For the same reason the mineral horizons below peaty topsoils are nearly always gleyed and single grained or massive. In peats and peat soils the term structure is more difficult to define since the plant fibres give the profile a different kind of structure, much influenced by the degree of humification. Well humified peat may be described as **amorphous** while non-humified or slightly humified peat may be broadly classified as **fibrous**. As expected from a structural assessment, these terms have implications for profile permeability. Well humified, cultivated peat soils (often described as fen soils or muck soils) can develop true soil structures. In the dry season the topsoil becomes a friable mass of loose crumbs. In drier climates the basin peats that have been drained and humified can develop an angular blocky structure and the ped faces become coated with iron ochre. This type of profile is referred to as **drummy peat**. A general indication of the relative permeability of the different peat soils is given in Table 5.2.

Table 5.2   Permeability of organic horizons

Less permeable		More permeable
Mossy	Herbaceous	Woody
Sapric	Mesic	Fibric
Amorphous	Fibrous	Drummy

Peat shrinkage

Peat accumulations are the result of plant growth in a special kind of environment and cannot be considered a permanent feature of the landscape if the environment is changed. Attempts to reclaim peatland for more intensive agricultural use will tend to remove the barriers to plant tissue decay which were necessary for peat formation in the first place. The loss of moisture from the peat initiates the general wasting away of the peat profile which is clearly recognised as **peat shrinkage**. Loss of moisture due to drainage works or by natural causes has several effects.

- Fresh peat contains a large amount of water (in the mossy peats as much as 97 per cent of its weight) and any loss of water causes considerable loss of volume.
- Loss of water allows air to enter interstitial voids and causes **oxidation** of the peat fibres. The rate of humification is greatly increased, fibres are converted into a paste-like humus and the denser peat occupies less space. If aerobic conditions are maintained, in time, all of the organic matter can be oxidised to carbon dioxide and water.
- In springtime when cultivated peat soils dry out the topsoil becomes a mass of fine, sooty crumbs which are very light. Any strong wind occurring before crops can provide protection can erode much of the topsoil.

The combined effects of these factors causes considerable and continuing lowering of the surface. Direct measurements of the rate of shrinkage have been obtained from several sources. In 1851 an iron pipe was driven into the Holme Moss, in the English Fens, until the top was level with the peat surface and the base secured in the stable material below the peat. The pipe now stands more than 4 metres above the surface, representing a peat shrinkage of about 3.0 cm/year. Various other sources report losses of between 1 cm and 11 cm/year. Compared with the rate of peat accumulation, these are very fast rates indicating that cultivated peatlands have a very limited productive life.

Conserving organic soils

The best minerotrophic peats which are used for intensive arable and horticultural cropping are being lost most rapidly. Each year the surface sinks further and the margins of the peat soil retreat to reveal more of the less fertile, often fine textured mineral material below. Although the loss of peat soil is inevitable in the long term, much can be done to reduce the rate of loss. Good agricultural practices which can maintain mineral soils in a condition of high fertility could be used to preserve the peats – at least in the form of peaty or humose soils. The very damaging system of continuous arable cropping could be changed to a system of alternate husbandry in which the grass crops and grazing animals return organic material to the land. In specialised arable areas a measure more likely to be adopted is that of **soil mixing** by which the rate of oxidation is reduced by mixing the peat with clay. This is a traditional practice achieved by carting in clay to spread on the surface, but the only economic method now available for the shallower peats is to drag a special heavy cultivator through the soil to bring up the underlying mineral materials.

Profile permeability

The hydraulic conductivity of peat on most sites is determined by the degree of humification. Drummy structures to a significant depth in the subsoil can be expected only in the driest areas. Since the rate of humification increases when the peat is drained it follows that drainage, to some extent, is a self-defeating exercise. It has been shown that the hydraulic conductivity of peat decreases exponentially with time after drainage and increases only in drier regions where, after several years, the under-drainage dries the peat profile sufficiently to allow the formation of drummy structures. This improvement of profile hydraulic conductivity cannot be expected in wetter regions. In fertile or potentially fertile peatlands it is advisable to design drainage installations to cope with completely humified peat in all cases where it can be expected that drummy structures will not improve profile permeability. In less favourable areas some less expensive measures can be adopted. These are discussed in Chapter 13.

The value of peat drainage

On most sites peatland drainage is a difficult undertaking and careful assessment of likely costs and benefits is essential. Each situation must be considered on its merits but the following points may be of value.

Still active blanket mires are usually not worth improving. Most are situated in remote, bleak areas where successful reclamation is unlikely and, where successful, could produce nothing better than improved hill grazing for a short time. On areas of former blanket mire colonised by heather, a range of worthwhile improvements is possible but the limited returns must always be borne in mind.

These same reservations apply to raised mires still active and in an ombrotrophic stage. Very little can be done with continuously wet mossy peat. In a similar way no longer active raised mires may repay some limited improvement since, initially at least, there may be a good gradient for drainage.

All minerotrophic mires, including former raised mires cut back to the minerotrophic stage by fuel gathering, are of much greater potential for farming because of the minerals they contain and the higher dry matter content of the peat. The best areas are already improved and any that remain may well have some inherent difficulty that has prevented earlier exploitation. The greatest difficulty is the cost of the work because most basin mires do not have good drainage gradients and many require pump drainage. Existing arable farming in the English Fenlands is possible only with pump drainage schemes organised on a regional scale.

Many peatlands are designated nature reserves and are protected by law, while others, although not protected, could be the focus of considerable public protest if there was any intent to alter the habitat. Of all farm wildlife sites, peatlands are particularly affected by drainage works. Bearing in mind the delicate balance of a peatland ecology and the disappointing results often obtained when reclamation is attempted, it must be concluded that many sites should indeed be left undisturbed.

The special difficulties of peatland drainage

Most improvement projects will involve basin mires and the drainage problem in basin peat is a persistently high groundwater table. The basic solution is to lower the groundwater table and in isolated upland hollows this may be achieved by gravity flow if the retaining rim of the hollow can be breached. Elsewhere it is often necessary to resort to pump drainage.

Chapter 6 Topography and land drainage

The term **topography** refers to the shape of the land surface and its individual units or landforms that make up the landscape. The study of landforms and the processes which produce them is called **geomorphology**. Many aspects of geomorphology are yet to be fully explained but the relationship between topography and soil drainage status is relatively well understood.

Factors which shape the landscape

Landforms seem to be the results of the interactions of **process**, **structure** and **time**. Process includes all natural agencies of weathering, erosion, transport and deposition. Structure in this context is a geographical term describing the broad nature and arrangement of rock formations on which erosion surfaces form. It must not be confused with fractures in soil profiles or in rocks which improve permeability. Erosion processes may act on the surface over limitless spans of time. Rates of action or mode of action may be modified by earth movement or change of climate or erosion may terminate when the land surface dips below sea-level where deposition is the dominant process.

Process

Landforms presently observed are the results of several processes, some of which are no longer active in a particular area.

Gravity

Continuous weathering has created the mantle of rock waste that covers the land surface. This loose material may remain in position for a time but eventually it will be moved to another part of the earth's surface by transport. The more elevated the point of origin of rock waste above sea-level the sooner and more rapid will be its movement to another site. The overall effect of rock waste movement can be described as the **downhill path**, which correctly implies that gravity is a major factor in geomorphic processes.

Mass wasting

When gravity acts alone the resulting movement of rock waste is called **mass wasting** and depends on the strength of exposed rock formation. Solid rock formations largely resist gravity and can form vertical surfaces, but loose formations, when left unsupported, slump to form a slope with an angle of between 25° and 40° from horizontal. Particles tend to fall downslope until there is equilibrium between friction and gravity and the surface is said to be at its **angle of repose** for the conditions present. The pile of dislodged fragments that collect at the bottom of a cliff is called **scree** and it rests against the cliff with the exposed surface at the angle of repose.

The slowest form of mass wasting is called **creep** and the rate of movement can be very slow on the least sloping surfaces. Usually plant growth is fast enough to cover any signs of movement. A major cause of creep is the freeze/thaw cycle of temperature changes. When topsoil freezes it expands and raises the ground surface appreciably. Individual soil particles move upwards perpendicular to the surface during the freeze but return vertically downwards during the thaw, resulting in a net downslope movement. Similarly, wetting/drying cycles, soil cultivations, traffic and scraping or burrowing animals, particularly earthworms, all combine to ensure continuing downslope movement of soil particles.

When topsoil becomes saturated and rests on an impermeable subsoil or frozen subsoil the downslope movement may be more rapid. This kind of downslope flow is called solifluxion, as described in Chapter 3, and is most effective in permafrost areas because the summer melt can penetrate further each year to maintain the supply of slurry. More rapid still are **earthflows** and **mudflows** which occur when a saturated formation reaches its liquid limit. **Landslips** involve deeper layers over shorter distances but are also the result of a layer becoming fluid underneath the main movement. The most devastating gravitational movements are **avalanches**. Most of the major earth and rock slips and many minor events are initiated by an earth tremor. Others are the result of prolonged heavy rainfall.

Water flow erosion

All kinds of rock waste, particularly that which collects in valley bottoms as a result of mass wasting, sooner or later is moved elsewhere by flowing water. Landforms in humid regions, where surface flows of water are common, reflect this influence of erosion by moving water which has been the dominant force for most of geological time. In recent times some maritime temperate lands have experienced glacial erosion which has modified the basically water-moulded landscape. However, water erosion is now modifying the glacial landforms. Wind erosion is significant only in dry regions and affects humid regions only along sandy shorelines and some easily blown arable topsoil.

Rainfall is a major force in shaping the landscape. When rainfall is heavy the water tends to flow across the land surface and will carry loose fragments of soil with it. The rate of erosion depends on the intensity of rainfall, the slope of the surface and the stability of the surface materials. Severe storms can cause flow across the whole surface as **sheetflow** but more commonly the water flows in a series of temporary tiny streams or rills causing **rillflow**. These may coalesce downslope to become streamflow which is usually permanent in a well defined channel. Small volumes of water require a greater slope to achieve the same erosive powers than large volumes so that the gradients of stream and river channels tend to diminish from source to mouth. As channel erosion proceeds it cuts deeper into the landscape. Mass wasting and surface erosion of channel sides causes the whole valley to deepen and widen at the same time and a system of valleys evolves with each river bed graded from source to mouth and each tributary joining at the correct level. The rate of water flow erosion decreases with gradient and when the whole land surface is reduced to near sea-level all further downcutting stops.

Erosion by ice

Glacial erosion has modified some temperate landscapes. Ice flowed radially outwards from centres of accumulation and where pre-existing valleys trended in the same direction as moving ice they were scoured and over-deepened, leaving discordant side valleys with tributary streams now joining the main channel over cascades or waterfalls. Where over-deepened valleys reach the sea they form a fjorded coastline. Where deposition occurred any cross valleys in the path of ice flow have been filled by drift and lowland areas have a drift-covered surface with a rolling landscape of rounded hillocks and water-retaining hollows. Water channels are modifying these landforms once more.

Deposition

Sediments carried away by river flow are deposited on the sea bed, often dispersed over a wide area, but some rivers discharging into calm water may build up an area of coarse sediments at the mouth called a **delta** which increases in size and advances seawards. Similarly, in shallow river estuaries, extensive **mudflats** may form which are relatively easy to protect and reclaim.

Landforms created by mass wasting and water-flow erosion

Mass wasting tends to be the dominant force on elevated parts of the landscape and where slopes are gentle it takes the form of soil creep. This process tends to form and perpetuate surfaces which are convex. In lower parts of the landscape where water-flow erosion is dominant the slope diminishes downslope in response to the effects of water erosion, creating surfaces that are concave. These two forces acting together create what might be considered the typical humid region topography of all curving surfaces.

Structure
Differential weathering

As discussed in Chapter 3, rocks are not equally susceptible to weathering and those rock types that most resist decay come to form elevated parts of the landscape. A good example of topography created by differential rock decay is seen in the Insch Valley area of northeast Scotland where easily weathered basic igneous rock forming the low-lying valley bottom is flanked by hills of more resistant slate to the north and granite to the south.

Differential erosion

For water erosion to have maximum effect it must flow across the surface in the greatest possible volume. This is achieved where the rock waste is impermeable and all incident rainfall remains on the surface. The rate of erosion is reduced if a proportion of the rainfall is absorbed by permeable rock waste and underlying rocks. In regions where sedimentary successions are exposed with their bedding planes at an angle to the surface – a very common situation – clay formations occupy the valley bottoms while more permeable rocks like sand, sandstone, chalk and limestone form the higher ground between valleys. This is known as 'scarp and vale' topography and is found in many parts of the world. The concept of differential erosion explains why valleys tend to have clay soils – the clay vales – while the slopes above have permeable soils. This also has implications for the patterns of groundwater movement as discussed in Chapter 9.

The importance of time

The period of time that a process has acted on a surface has an important influence in topographical development. This depends on three factors. The first is that flowing water erodes more quickly on steep slopes than on gentle slopes and erosion ceases altogether when the whole surface is reduced to a plain near to sea-level. The second factor depends on relative rates of changes of surface elevation due to different forces; for by careful measurement it has been deduced and sometimes observed that crustal warping movements alter the surface elevation much faster than erosion processes, although both are barely perceptible in terms of any life-span. The third factor is that erosion continues with little interruption while crustal movements are erratic, tending to be active for short periods then remaining inactive for much longer periods in a particular region. Consider a low-lying plain or an area of sea-floor, either of which may be elevated by crustal warping to form a plateau well above sea-level. Rivers flowing off the edge of the plateau surface will cut narrow, steep-sided valleys across the raised surface because of the accelerated rate of river erosion. This type of landscape with narrow valleys separated by wide plateau surfaces is classified as **young** or **immature**. As river erosion and mass wasting proceed the valleys widen and the plateau is reduced until a rolling landscape of hills and valleys is created in which there are very few areas of flat surface. This is called a **mature** topography. Continuing erosion widens the valleys and reduces the hills until, in time, the whole surface is reduced to a plain at or near sea-level with only minor undulations above the general surface. Such topography is described as **old**. The progression of landscape development with time is shown in Fig. 6.1. This is a theoretical concept but all such landscapes can be observed around the world and their influence on land drainage is valid however they are formed.

Basic landform shapes

The landscape can be divided into a small number of basic shapes or 'building blocks'. These shapes are flat surfaces, both horizontal and inclined, and four types of curved surfaces which are illustrated in Fig. 6.2.

Figure 6.1   Landscape development

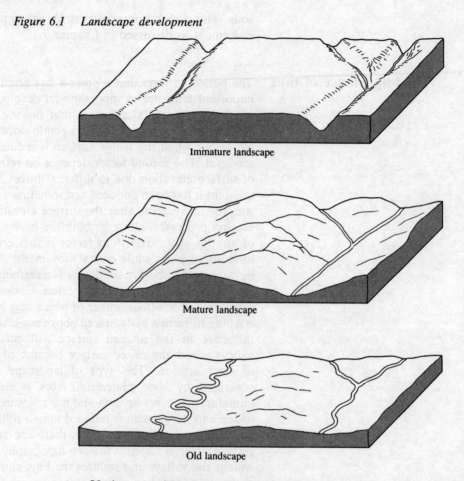

Immature landscape

Mature landscape

Old landscape

Various combinations of the basic slopes can form any of the observed landforms. Four units of shape 1 topped by four of shape 2 will form a water-retaining hollow while four units of shape 3 with four units of shape 4 above them will represent an isolated hill. The nature of the surface slope greatly influences the drainage characteristics of the soil. Water falling on an impermeable level plain will tend to accumulate on the surface until there is sufficient water present to cause lateral flow which should be radially outwards, but truly horizontal surfaces that might produce this effect are very rare. Water falling on an inclined plain will tend to flow across the surface as a series of parallel rills that do not disperse or concentrate and the surface may be

Figure 6.2   Basic landscape units

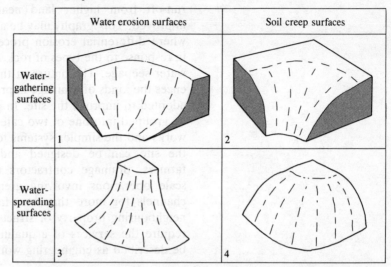

described as **neutral** in terms of water flow. Rain falling on surfaces like shapes 3 and 4 of Fig. 6.2 will tend to flow along diverging lines so that slopes composed of these basic shapes are **water-spreading surfaces**. Water flowing across a surface composed of shapes 1 and 2 will tend to converge towards a point on the surface so that the landform is a **water-gathering surface** and where the area is large enough there will be the source of a stream at the point of concentration at the bottom of the slope.

The relationship between topography and soil drainage problems

Water is influenced by gravity and a given volume will flow down a sloping surface at a velocity that is proportional to the **gradient**. The gradient of a slope is variously expressed as a number of degrees from horizontal, as a percentage or as a ratio. For example, a fall in the surface of 1 metre every 100 metres is a gradient of 0.6° or a 1 per cent gradient or a 1 : 100 gradient. Drainage designers must be very aware of surface gradients because they have a major influence on the feasibility and costs of a chosen design. The landforms present, both within and around a site to be improved, influence the nature of the drainage problem he has to solve. The topography may be the direct cause of the poor drainage status where it restricts surface drainage to the extent of causing a high groundwater table or where surface run-off from higher land nearby collects in the site to be improved. Topography may be an indirect cause of poor drainage where differential erosion processes have shaped the landscape in response to the types of rock, so influencing the sites of spring-water seepage. The nature of the topography also strongly influences the kinds of drainage improvement techniques that must be adopted to improve the site. In general, land drainage works can be grouped into one of two categories. In-field soil improvement works and the simpler systems for removing drainage water from the site can be designed and installed quite adequately by farmers, drainage contractors or agricultural advisers. Larger scale operations involving greater volumes of water, outflow channels for more than one farm, public safety, safeguarding neighbouring property or structural strength calculations usually require the services of a qualified engineer. Such facilities may be described as engineering works.

Draining mature landscapes

Drainage works on mature landscapes present few engineering problems since suitable outflow routes and surface gradients are easily obtained. Some special techniques may be needed to avoid steep gradients or to control the very fast flows which they generate. Slopes with gradients of 15 per cent or more usually do not require drainage improvement since excess soil water is easily shed downslope. In any case, such sites are difficult to drain because machines cannot operate and because soil creep or earth slips may disrupt drainage installations. Elsewhere the normal range of in-field drainage problems and solutions can be expected. On water-gathering sites the soil receives more water than falls as direct rainfall and for certain soil types waterlogging occurs. No special technique is needed to drain the land but allowance should be made for the greater effective rainfall. The concentration of surface and subsurface lateral flow will increase towards the point of focus of the slope. On the other hand, on water-spreading slopes the soil has to absorb less water than is provided by rainfall and soil saturation is delayed. Such sites will normally require a less intensive drainage system. If the slopes appear to be wetter than would be reasonably expected from the soil type and amount of recent rainfall then a spring seepage should be suspected. Drainage design should take account of the varied topography and, where large areas are to be drained, it is important to divide the area notionally into smaller areas each with a more uniform effect on surface water flow. An appropriate drainage design may then be applied to each subdivision of the area.

Draining old landscapes

An old landscape, or any wide expanse of relatively level, low-lying countryside, whatever its origin, usually presents engineering problems. Basic difficulties are the lack of gradient to allow sufficient gravity outflow from drainage installations. As a last resort it may be necessary to pump the water out to achieve satisfactory clearance of excess water. Special provision may be needed to prevent the entry of water from nearby land and defence works may be needed for protection against flooded rivers or high tides. In the most difficult areas, productive agriculture is possible only if land drainage is organised on a regional scale. Difficulties may remain in the field after the engineering problems have been solved. There is little freedom of choice of drainage works and high standards of accuracy are needed at the installation stage.

Draining young landscapes

Drainage work on a young landscape is intermediate in character between that needed for mature and old land surfaces. The valley sides will be too steep for drainage improvement and in general the valley bottoms will be too narrow for cultivation. In the wide plateau areas between the valleys there may well be engineering difficulties for areas of wet land remote from the valleys. However, many young landscapes have plateaus with a general tilt or a surface with gentle slopes that are relics of an earlier cycle of erosion. In either situation drainage difficulties are much reduced.

Draining glaciated landscapes

On glaciated landscapes, which are young in terms of water flow erosion, the most common feature of significance for land drainage is the water-retaining hollow. The depressions may contain free standing water or they may be completely filled with peat. The special features of draining peat-filled hollows are discussed in Chapter 15. Some further difficulties may be encountered when cutting an outflow channel through the retaining rim of the hollow. Very deep excavations in drift can be dangerous, requiring supports for the side walls, and the excavation may have to be done in steps where the required channel depth is beyond the reach of the excavator. If solid rock is encountered the services of an explosives expert may be required. Not all water-retaining hollows are clearly evident at the surfaces. It is sometimes found that glacial erosion has formed a hollow in the solid rocks or on the surface of the glacial till but a later phase of glacial activity has completely filled and buried the hollow with sand and gravel.

Where the site is level a field survey provides the rather puzzling evidence of a high groundwater table in one part of the area and free drainage elsewhere. An equally puzzling variation occurs where a buried hollow is sited near the upper margin of a steep bank and the porous soil would be expected to be freely drained, if not overdrained. Exploration pits at the margins of the wet area will reveal the solid rock or bank of impervious till projecting to near surface level. Many hollows can be completely drained by doing nothing more than cutting a channel through the rim.

High groundwater tables

Wherever natural drainage is restricted by the nature of the local or regional topography the soil drainage problem most likely to be encountered is that of a persistently high groundwater table at, just below or above the ground surface. The zone of saturation may rise and fall in response to seasonal weather changes but its presence high in a soil profile will severely limit soil fertility. The basic drainage need is to lower the groundwater table. The natural level of a groundwater table is a function of topography and is *not* influenced by the type of soil, rockwaste or solid rocks present. Where impermeable soil or rock formations occupy an area of restricted drainage the effect of a high groundwater is seen in the form of ponded water or of channels standing full of water. The groundwater is effectively displaced by impermeable material occupying the ground surface and a profile pit would indicate the effects of a perched watertable.

Chapter 7 **Rainfall**

The occurrence and amounts of rainfall vary over the land surface for a number of reasons. Such variations are important for agriculture and the prediction of rainfall amounts is an essential feature of production planning. Methods of statistical analysis can be used to assess likely rainfall amounts and these are the basis of calculating the amount of excess rainfall that must be removed from the soil by land drainage installations.

Air masses

The general climatic characteristics of any land area are determined by its latitude and proximity to an ocean but the day-to-day variations of weather result from the properties of the different air masses that can cross the area. Air masses lying over particular regions of the earth's surface tend to acquire characteristic properties and become distinctly homogeneous. Air lying over a region of tropical ocean becomes heated, absorbs large amounts of water vapour from the ocean and is called **maritime tropical air**. Over the cold, high latitude oceans **maritime polar air** absorbs less water vapour because it is cooler but is relatively moist compared to **continental tropical air** and **continental polar air** which originate over land surfaces. These distinctive air masses may be moved from their source areas by atmospheric circulation and dominate the weather along their routes until their distinctive characteristics are lost. Most of the rain falling on land surfaces is derived from maritime tropical air.

Rainfall

As maritime tropical air moves over the surface any factor that makes it rise also causes cooling and reduces its capacity to retain water. Excess moisture condenses out, forms cloud and may fall as rain. There are three main reasons why air may be forced to rise.

1 **Cyclonic rain** occurs within great swirls of the atmosphere called cyclonic depressions where maritime tropical air is forced to rise over maritime polar air. The shed water vapour causes widespread cloud cover and prolonged rainfall.

2 **Orographic rain** occurs where an air mass rises to cross a mountain barrier. The rising, cooling air sheds considerable rainfall on the windward side of the mountains. On the leeward side of the barrier, however, the air is descending to its original level and becomes warmer as it descends so that no moisture is shed. The drier region on the lee side of mountains is called a **rain shadow area**.

3 **Convectional rain** results from solar heating of the land surface. In hot weather the land surface warms the lowest layer of the atmosphere, causing it to rise as a series of pockets which are seen as developing cumulus clouds. If surface heating is severe and the distribution of atmospheric temperature with altitude is such that the rising cooling pockets of air remain warmer than the surrounding air up to a great height then strong up-currents can form causing thunderstorms and heavy local rainfall.

Climatic zones   The world climatic zones are determined by latitude (distance from the equator) and by the spin of the earth. This potentially regular arrangement of parallel climatic zones is disrupted to some extent by the positions of oceanic and continental areas which alter the atmospheric pressure distribution patterns seasonally and consequently change the routes of the air masses. The combined effects of these factors determines the regional climatic variations observed in temperate latitudes. There is a strong relationship between regional climate, soil properties and the benefits to be obtained from land drainage.

Maritime regions   Mid-latitude land areas that lie in the path of cyclonic depressions and receive appreciable amounts of rainfall all year round have a **maritime** environment. Western Europe, most of eastern North America and the coastlands of Western Canada and north-west USA are maritime regions. Rainfall is evenly distributed throughout the year, winters are relatively mild and summers tend to be cool. Rainfall is mainly cyclonic and falls in greatest amounts near coasts where 1 000 mm per year is common. It decreases inland to about 500 mm which is taken as the limit of a maritime environment. In front of mountain barriers orographic rain increases total rainfall in lowland areas to about 2 000 mm and in some high mountain passes it may exceed 4 000 mm year. In the wettest areas there is a distinct peak of rainfall in the winter months. Potential evapotranspiration ranges from zero in bleak northern and coastal areas to about 500 mm year in milder southern and inland margins. The effects of a continental climate increase towards the limits of inland maritime regions. A smaller annual rainfall tends to be concentrated more in the summer months due to convectional rainfall and winters are colder and drier.

Brown forest soils are found in all drier and warmer parts of maritime regions. Soils are likely to be at field capacity or wetter in winter and an appreciable soil water deficit can be expected in most summers. A wide range of crops can be grown and drainage works are usually worthwhile. Much summer rainfall is absorbed by the soil profile to reduce the soil water deficit but most winter rainfall becomes excess soil water. Drainage works to remove excess water will protect soil structure, permit winter-sown crops, accelerate soil drying in spring and improve trafficability for harvesting operations. **Podzolic soils** occur in wetter, cooler areas. Total soil saturation is common in winter and the summer soil water deficit is not great, often absent in poorer years. Grassland farming predominates with some crops in more favoured districts. Drainage works can reduce soil poaching and extends the grazing season on better pastures. **Gley soils** in any environment will need drainage improvement works if they are to become productive.

Continental regions

Each temperate land mass has an interior **continental** region. Low atmospheric pressure in summer allows penetration of cyclonic depressions but high pressure in winter diverts depressions away from central regions. Winters are intensely cold and summers are very hot. What rainfall there is occurs in the summer months and is often caused by convectional forces. Rainfall varies from about 500 mm in the grasslands to very little in desert areas. Potential evapotranspiration throughout the region is at least equal to rainfall. **Prairie soils** are dormant in winter and receive only moderate amounts of rain in summer. Drainage works are not likely to be required. Climatic zones have transitional margins with **transitional soils** and one such zone of great importance to agriculture is the boundary between forest soils and prairie soils. These broad zones include much of central and eastern Europe and the eastern margins of the central plains of North America. The maritime aspect of the climate permits the growth of a wide range of valuable cash crops like soya and maize but continental influences can cause severe convectional rainfall which can damage these crops. Drainage works may be needed to remove excess soil water before soil saturation can inhibit crop growth and reduce yields.

Other climatic regions

Asia has an eastern **monsoon zone**. The high pressure of the continental interior in winter causes dry cold air streams to arrive from a northerly quarter. Onshore winds from an easterly direction bring warm, often very wet conditions in summer. Both continental land masses in the northern hemisphere have a south western **mediterranean** climatic zone. The mediterranean lands and California have such a climate. Cyclonic depressions can cross the region in the winter months but are kept out by a northern advance of the tropical high pressure zone in summer. Winters are warm and moist while the summers are hot and dry. Crops are sown in autumn and harvested in spring. Continental, monsoon and mediterranean climates have in common an inhospitable season when the land is barren, followed by a short intense growing season. Land drainage works to remove excess soil water are not often needed in a mediterranean climate. However, monsoon lands are associated with rice growing, which requires a precise control of flood water.

Soil drainage requirements

Where excess soil water can remain in the soil long enough to inhibit crop growth it is necessary to consider some type of land drainage installation. These must be suited to the drainage problem and have sufficient capacity to clear excess soil water in reasonable time. On the other hand, they must be as cheap as possible. To get the best balance it is necessary to take account of the rainfall regime of the area. Rainfall is very variable and it would not be practical, in most cases, to cater for the most severe storm events, but it is necessary to accept that, on occasions, the drainage system will be overloaded. Drainage design is concerned with choosing the optimum capacity for drainage installations. **Excess soil water should be removed at the minimum acceptable rate**.

The critical season   Soils in the different climatic regions have drainage problems only in the **critical season** when any excess of soil water can damage the agricultural economy. In the maritime arable cropping areas, with rain all year round, the critical season is the whole year but any drainage works which can cater for the winter season will also be satisfactory for the rest of the year. In the exposed grassland areas of the maritime regions little improvement can be expected in the winter season but the summer grazing period can be improved and extended by land drainage – the critical season being spring, summer and autumn. The rainfall peak that often occurs in the winter season need not affect the choice of design capacity. In most other regions the critical season is the six-month period when the crops are grown whether it be in summer or winter. For erosion control the critical season is the non-cropping part of the year when the land is bare.

Rainfall amounts   Periods of rainfall vary in duration, intensity and in areal extent. Where sufficient records are available the data can be examined statistically. The information, obtained by statistical analysis, is of great value for drainage design and the characteristics of the rainfall regime should be the basis of calculations of the volume of water to be removed. Analysis of rainfall amounts shows that the more a period of rain exceeds the average rainfall the less often it is likely to occur. Thus a period of rainfall of a particular amount will occur in any area at intervals of time which are related to the amount by which the rainfall exceeds the 'norm'. This interval of time is called the **return frequency** of a given amount of rainfall. The amount of water to be cleared from the land and, therefore, the cost of the work are related to the return frequency selected as most appropriate for local conditions. Rainfall events with return frequencies of once per year or once in 2 years are modest amounts, once in 5 years is a moderate storm event, while once in 10 years is a noteworthy storm event. Events occurring once in 100 years or more would be considered as natural disasters. With statistical analysis, the term 'once in 5 years' does not strictly mean that each fifth year will have a storm of a particular value but it can be assumed with reasonable accuracy that such storms will occur something like twenty times every 100 years.

Choice of return frequency

The choice of an appropriate return frequency for drainage design purposes is determined by the climate and economic considerations since it is necessary to balance the capital costs of protection against the likely increase in returns resulting from the investment. In this context only agricultural crop losses are considered. Where lives and property are at risk and long-term return frequencies are involved the work must be left to qualified engineers. For farming purposes the amount of damage that is acceptable depends on the nature and value of the crops at risk. To strike a balance the following criteria can be used:

- For very high value crops of a specialist type that are sensitive to waterlogging of the root zone it may be necessary to design for once in 25 years return frequencies.
- For high value horticultural crops and the more valuable agricultural root crops like sugar beet and potatoes or crops that are sensitive to waterlogging, like peas, a return frequency of once in 10 years is advisable.
- For less valuable root crops and specialised arable farms growing mainly winter-sown cereals a return frequency of once in 5 years is satisfactory.
- For mixed farms with spring-sown cereals and grass a return frequency of once in 2 years is satisfactory unless root crops are included in the rotation.
- Most species of grass can tolerate periods of flooding and the returns from draining grassland are indirect, being measured as extra days of grazing and less surface damage. For grassland farms in the wetter areas a return frequency of once per year can be adequate.

Rainfall intensity

Analysis of rainfall data also shows that heavy rainfall lasts for a shorter time than more moderate rainfall events. The amount of rain falling in a given time is called the **rainfall intensity** and is expressed as millimetres/hour, day or month as required. Examination of rainfall records shows that the average amount of rain falling each day of a storm period reduces as the length of time being considered increases. A search for single-day rainfall figures will reveal a number of days with very high rainfall values. If the period is increased to any 5 consecutive days the average daily falls for peak periods will always be less than single-day peak falls. Similarly, average falls in peak 10-day rainfall periods will be less than those for 5 days. This is important for drainage design work. If a design involves consideration of a 1-day rainfall intensity with a return frequency of, say, once in 5 years the amount of rainfall to be removed can be quite high. The volume of water to be removed in any 5-day period of the same return frequency is much less and drainage costs are lower. For most types of farming a short period of waterlogging does little or no damage so designs based on 5-day periods are quite satisfactory. Choosing such rainfall values may allow the land to be saturated for one or two days but the drainage system will be able to cope with the total fall of any 5 days before any damage occurs. However, certain crops and land drainage improvement work which includes soil treatment often cannot tolerate any flooding in which case a 1-day storm value must be selected as a basis of drainage capacity.

The area affected by rainfall events

Rainfall data also show that heavy rainfall occurs over a smaller area than more moderate rainfall. The most intense storms are usually convectional and restricted in areal extent while the gentler rains of cyclonic depressions extend over whole regions. Precipitation values for a single weather station may not always truly reflect the actual precipitation over a wide area, particularly if taken over a small time period. Rainfall values should be derived from as long a period as possible and from as many recording stations as are available.

The design rainfall rate

The amount of rainfall that must be cleared from the land is called the **design rainfall rate** and is measured as millimetres per day in the same way as rainfall intensity. The design rainfall rate can be established using the following steps.

1 *Determine the length of the critical season*. The critical season in maritime regions will usually be the whole year but in the very wet areas where there is a peak of rainfall in the winter months some savings in costs are possible by choosing a 9-month critical season based on spring, summer and autumn rainfalls. The stock are usually taken off the best grass in the winter season. For other climatic regimes the 6-month growing season should usually be the basis of calculation.

2 *Establish the average daily rainfall of the critical season*. This requires the best rainfall data available. Where weather stations are remote or widely separated, some subjective adjustments may be required to allow for local effects like orographic rainfall or a rain shadow. For major improvement works it may be necessary to delay the design work until a climatic survey has been completed.

3 *Choose a suitable return frequency*. This is based on the type of crops to be grown as described above.

4 *Select the appropriate duration of rainfall*. Select either 1-day or 5-day storm events. This depends on the susceptibility of the crop to waterlogging of the root zone and whether the drainage works selected can be allowed to be overloaded. Calculate the smallest acceptable design rainfall rate.

Some countries (e.g. the United Kingdom) have all such rainfall values already calculated and presented as climatic regions for drainage design purposes. Where available this type of information should be used. In other areas the guidelines suggested in Table 7.1 can be used unless local knowledge indicates otherwise. The method can be demonstrated by considering an all-grassland farm with an average daily rainfall during the critical season of 2.5 mm. The low-risk crops can tolerate some soil saturation. In this case a once-per-year, 5-day rainfall event would require a design rainfall rate of 2.5×5 – say, 13 mm/day. For many sites in arable areas, design rainfall rates fall mostly within

Table 7.1

Design rainfall rate (mm per day)
=
Average daily rainfall of critical season (mm per day)
×
Factor (*F*)

Value of *F*

Return Frequency	Once per year	Once in 2 years	Once in 5 years	Once in 10 years	Once in 25 years
5-day storm period	5.0	5.5	6.0	6.5	7.0
1-day storm period	11.0	12.0	13.0	14.0	15.0

the range of 7–13 mm/day. Where drainage work in a high rainfall area includes soil treatment and 1-day events must used, the design rainfall rate can be much higher.

Acceptability of the method

This rule-of-thumb system for drainage design can be used with confidence for the great majority of in-field land drainage works and for associated surface outflow channels serving land areas up to about 40 ha of agricultural land. For more complex cases involving other kinds of land surfaces and paved areas and for larger areas of land which may provide a sufficient volume of water to flood nearby property, it is better to use more precise methods of assessing outflows. In such cases it is advisable to consult an experienced engineer.

Chapter 8 **Surface water**

In maritime and other humid regions, excess rainfall flows across the land surface and where there is sufficient flow, permanent channels are formed. This natural surface drainage pattern consists of all sizes of channels from tiny runnels on field surfaces to the widest of rivers. The whole drainage system so formed is not static but seeks to achieve equilibrium with the rest of an ever-changing environment. By variously eroding, transporting and depositing they adjust to prevailing conditions. Open channels constructed to aid land drainage become extensions of natural drainage systems and tend to obey the same laws. Both types of channel are important for land drainage by acting as **outlets** for excess soil water collected by in-field drainage systems. The most relevant properties of flowing water are discharge, velocity, erosion, transport and deposition.

Channel discharge   The volume of water passing a fixed point in a watercourse in a given time is called the channel **discharge** at that point.

The discharge may be expressed as $Q = AV$
Where Q = the discharge in cumecs
A = the cross-sectional area of water in m^2
$$= \frac{\text{top width} + \text{bottom width}}{2} \times \text{average depth}$$
V = average velocity of flow in m/s

It is expressed in m^3/sec (cumecs) for larger flows and litres per second (l/sec) for smaller flows as appropriate (1 m^3 = 1 000 litres).

The dimensions may be measured directly and the flow rate deduced by timing a floating object between measured points. The timed object should be floating at midstream to obtain the most accurate results. More precise measurements of flow velocity may be obtained using a water-current meter. Where regular monitoring of a stream discharge is required it is more convenient to construct a weir across the flow so that the whole discharge must pass through a sharp edge, triangular or rectangular notch in the weir. Graduated markings on the side of the notch are equivalent to known discharge values allowing immediate assessment.

Run-off

The amount of water shed by a given area of land in a given time is called the **run-off** and can be expressed in convenient terms like cumecs per square kilometre. A natural water-collecting area is called a **catchment** with the gathered water leaving a particular catchment as a single stream, the size of which is determined by the run-off. Assessment of run-off is an important part of drainage work and for this purpose it is very useful to be able to visualise the amount of water collected by a known area in the locality. A stream should be chosen and the discharge estimated at a suitable point. The catchment area supplying this amount of water can be found by using a 1 : 50 000 scale map of the area and some tracing paper. Note on the map where the discharge was measured and then trace the line of the stream and any feeder tributaries upstream until all the supplying channels are included. Trace in the ridges of higher land and other areas without surface channels which act as **watersheds** and delimit the study catchment from adjacent catchments. The area of the study catchment is then found by counting the grid squares on the map (1 km² each), making allowance for squares partly included. If opportunity permits this should be repeated for different flow conditions at the same point and for different catchments with a range of sizes.

Channel dimensions and flow velocity

Observations of many streams and rivers of all sizes in humid regions have shown that flowing water cuts a channel into the land surface in a precise manner and that the relationship between discharge, stream width, stream depth and flow velocity are by no means haphazard. As a given watercourse changes from a condition of low water to full flood (bank full to the ground surface) measurements at a point in the channel show that water-flow depth, width and velocity all increase at rates which are related to the discharge. As the discharge increases and decreases the other factors also increase and decrease in proportional amounts (as small, positive, exponential functions of discharge). Furthermore, when measurements are taken at successive points downstream on a single watercourse (when the water is reasonably steady) it is found that these same constant relationships prevail. Thus, as the discharge increases from channel origin to outlet, the width, depth and flow velocity all tend to increase by inter-related amounts. In circumstances where one factor is inhibited the other factors tend to adjust in compensation. For example, where channel width is constricted through a narrow gorge cut in hard rocks or by bridge piers the uniform discharge is maintained by increased depth and flow velocity. In a similar manner, where new channels are excavated, for a given discharge, a wide channel has a lesser flow velocity than a narrow channel and the same natural forces produce a greater flow velocity when channels are shortened and straightened.

Other factors influencing flow velocity

The velocity of water flowing in a channel can be attributed to several factors of which channel gradient is perhaps the most obvious. Velocity of flow is influenced just as much by the **hydraulic gradient** which is formed whenever the surface of a body of water is not precisely horizontal. Water will seek to flow in any direction that will re-establish a horizontal surface. The difference in height between one part of the water surface and another is called the **hydraulic head** and this is an important factor for all land drainage, water supply and irrigation works. The greater the head, the greater the hydraulic gradient and the greater is the resulting flow velocity. In pipes or even in channels deep enough to contain the water, hydraulic gradient acting against the channel gradient may cause water to flow upslope. Where hydraulic gradient acts in the same direction as channel gradient the velocity of flow is greatly increased. This explains why a narrow point in a channel increases depth and velocity. Water is dammed up behind the constriction and the hydraulic gradient is increased.

Another factor influencing velocity is **channel rugosity**, which is a measure of the roughness of the surface over which water must flow. Rugosity has more influence on small discharges because a larger proportion of water comes in contact with the channel surfaces. For the same reason a wide, shallow channel permits a lesser flow velocity than a narrow, deeper one for the same discharge.

All of these factors help to explain channel flow characteristics. Velocity of flow tends to increase with discharge along a natural channel from source to outlet because of reduced friction and increased hydraulic gradient even though the channel gradient decreases. This often comes as a surprise to casual observers because tiny flows appear to move quickly over rocks in the channel bed while deep rivers appear to move sluggishly. It may be said that large volumes of water flow more efficiently than small volumes.

The relationship between rainfall and channel discharge

As already discussed, incident rainfall may either infiltrate the soil profile and become groundwater or it may flow over the surface as run-off. Clearly, surface channels must be affected directly by run-off. In a maritime region maximum discharges will occur when catchment soils are saturated in the winter half of the year and will be least in dry spells in the summer half of the year. In small catchments the channels may dry up when soils in catchment areas have a soil water deficit. In larger catchments there is enough groundwater entering the channel to ensure continuous flow and this minimum amount is called the **base flow**.

For drainage works it is more important to evaluate **peak flow** discharges which are the maximum likely values for storm periods of a chosen return frequency. Peak flows depend not only on the duration and intensity of rainfall but also on the nature of the catchment. For larger catchments the calculations can be complex. Important factors are the permeability of the catchment surface, the degree of soil saturation before the storm begins, the size and gradient of the catchment and the shape of the catchment in relation to the prevalent direction of storm movement. Catchments with several tributary streams may deliver flood flows to the main channel in succession to provide a lengthy period of moderate flood discharge values or all tributaries may deliver all flood waters at the same time to form a severe flood of shorter duration. Much depends on the **time of concentration** of flood water. Similarly, in long narrow catchments much depends on the alignment of the main axis in relation to storm movements. A storm crossing the main axis causes only a small peak discharge.

A storm moving upslope along the main axis may cause a moderate discharge of long duration because the channel tends to clear itself as the storms move along. When a storm travels downslope along the main axis its flood waters from the head of the catchment may keep pace causing a severe flood when the outlet channel is reached. Some catchments are particularly prone to **flash floods** which are a rapid build up to peak discharges after the commencement of a storm. Most of these have steep gradients and other properties which ensure a short time of concentration. The nature of catchments and storms is a local knowledge factor which must be considered at an early stage of design work.

Channel erosion
As the water moves along its channel, loose particles are dislodged from the bed and sides and are carried along by the flow. Larger fragments continually scour and chip the channel so that even the hardest rocks can be worn down in geological time scales. The bombarding pebbles become rounded in the process as they are moved along the channel. All material being moved along with the water is called the **stream load**. This is composed of three categories:

1 The **dissolved load** consists of materials dissolved in the water and generally includes ions of sodium, calcium, potassium, carbonate and nitrate removed from rocks and soil.
2 The **suspended load** consists of the smaller rock waste fragments carried along in suspension by the turbulence of the water.
3 The **bed load** is formed by the larger rock fragments that are moved along the bed by sliding, rolling or by a series of hops.

The size of fragments that can be transported depends on the velocity of flow. Still pools only carry the dissolved load while rivers in flood can move large boulders. The suspended load increases with discharge to the power of between 2 and 3. That is, if the discharge increases threefold then the suspended load will increase to some value between 9 and 27 times the original value. The bed load is difficult to measure but it appears to behave in the same manner. Any small increase in discharge increases the stream velocity and allows it to carry a larger load.

There is also spare energy in the flow to further erode the containing channel.

Conversely, any decrease in velocity – as a flood subsides or where a section of reduced channel gradient occurs – causes a loss of transport capacity and deposition results as part of the load falls out of the flow and comes to rest in the channel. The largest rock fragments drop out first followed by progressively smaller particles as velocity decreases. Erosion in fast-flowing sections and deposition in slow-flowing sections of a channel provides a mechanism for streams and rivers to grade and smooth out the channel bed on which they flow.

Base levels

Streams and rivers with sufficient gradient cut their own channels steadily downwards into the rock formations. The channel may migrate laterally where erosion picks out a softer seam of rock but generally, once a channel has developed, it tends not to move as long as down-cutting is the dominant process. In each river system the limit of all downwards erosion is sea-level at the mouth where all gradient is lost. Sea-level, therefore, is the **base level** of any river below which water flow erosion is impossible. Rivers flowing over old landscapes have smoothly graded profiles with channel gradients reducing steadily from source to base level at the outlet. Rivers flowing across a young landscape may have to cross rock outcrops that are markedly harder than the others. The most common are igneous dykes and sills and these rocks may resist water flow erosion sufficiently to form a step in the river channel where rapids or even a waterfall can occur. Such rock barriers form **local base levels** which act as the limit of downwards erosion in the section of channel immediately upslope. Given sufficient time, as the landscape ages, all steps are slowly smoothed out as the harder rocks are worn away.

River meanders and flood plains

Because downwards erosion is limited by the presence of a base level in the channel, an ever-lengthening section with a flat gradient is formed on the upstream side of the barrier forming the base level. But water flowing across the level section still possesses considerable kinetic energy which becomes available for sideways erosion. Any chance irregularity in the channel can deflect the main flow towards one of the banks where increased erosion can cause the development of a small embayment. The water then flows round the longer path and is accelerated, further increasing outwards erosion. At the same time flow is reduced at the opposite bank where deposition exceeds erosion and a bar of sand or gravel gradually extends into the channel. Furthermore, when the main flow leaves the bend it is then directed towards the opposite bank further downstream where another bend develops. In time the whole channel becomes a series of loops or **meanders**. Each meander automatically directs the main flow towards the outside bank, causing the channel to follow an ever longer and more convoluted path. But a controlling mechanism puts an outer limit on the growth of loops. As the loops get bigger they encroach on each other and flood flows may burst through from one loop to another or elaborate loops may pinch themselves off at the neck.

Both events leave a loop of channel no longer used by the main flow. This is called an **abandoned meander** and remains as a marshy, horse-shoe-shaped hollow cut off from the main channel. Because the erosion always tends to be greater on the downvalley side of each meander loop, the loops slowly migrate down the valley as units and the valley is subjected to the passage of a series of loops over a long period of time. The outer extremity of each loop may erode the valley sides, widening the valley bottom and forming a flat area called a **flood plain**. The limits of the flood plain are marked by more or less steeply sloping banks called the **marginal bluffs**. The maximum loop size is proportional to channel width so that the width of the flood plain is also a function of channel discharge. Periods of flood may further erode the marginal bluffs and any permanent increase in discharge resulting from change of climate or river capture will also cause widening of the flood plain. The widest flood plains are formed by the widest rivers at their broadest point near the sea.

Braided streams and rivers

The patterns of erosion and deposition discussed above occur mainly in humid regions where discharge is continuous and relatively steady. In arid or semi-arid regions surface channels flow only in the short periods of usually intense rainfall. In the dry season the channels may become partially blocked by wind-blown sediments and the sudden torrential flows are heavily loaded with rock waste. In these circumstances a **braided channel** develops with the main flow split into smaller flows by a pattern of shifting, changing islands of sand or gravel. Braided channels can occur wherever the discharge of the wet season is markedly greater than that of the dry season and there is a plentiful supply of rock waste. In such environments deposition may exceed erosion and the coarser sediments may slowly accumulate to fill the valley.

The effects of glaciation

The process of valley erosion is further complicated by changes of sea-level. During major glaciations when the hydrologic cycle was interrupted by the accumulation of ice on land surfaces the sea-level dropped by as much as 40 m below its present level, providing rivers with a different base level. Beyond areas of ice cover all valleys were cut well below the present base level and in many the seaward end was drowned by the returning sea when the ice melted. They are now seen as the familiar wide **river estuaries**. Beneath ice sheets existing valleys could be deepened far below sea-level by ice erosion, as at Loch Ness in Scotland, or they could be filled by glacial drift like parts of the original channel of the Clyde. When the ice melted, torrential flows could erode **meltwater channels**, some of which have become routes for diverted streams.

A feature of glacial decay is the development of braided streams which tended to fill outflow channels with complex bedded sand and gravel layers. There was no discharge during the winter freeze and waste-laden flows during the summer melt period. Many rivers in formerly glaciated landscapes now flow over these varied but unconsolidated glacial and post-glacial deposits where erosion can be relatively rapid. However, many have bedrock projections which cause frequent steps in an essentially young river profile.

River flood plains

River flood plains, as the name suggests, become flooded each time the river is swollen by heavy rainfall. The muddy flood water can fill the whole area between the marginal bluffs and the whole system acts as a reservoir for the excess water. Where the land in the flood plain is cultivated the storm flows can cause damage but in a natural community the plants are little affected. The flood water is slowed and filtered by the vegetation and a film of mud is left each time the water subsides. A succession of floods over many centuries has built up a layer of medium-and fine-textured particles called **alluvium** which is often about 1 m deep in formerly glaciated lands. Poorly drained hollows and abandoned meanders in the valley bottom may cause peat accumulation from time to time as the flood plain surface changes and it is not uncommon to find lenses of peat or peaty alluvium in the profile. Any downcutting by the main channel will usually involve a lowering of the flood plain at the same time but, in some situations, remnants of earlier, higher flood plains may be left behind and these more elevated steps in the valley bottom are described as **river terraces**. These terraces are well above the groundwater table and are much prized for intensive cultivation.

Valley cross-sections

All of these processes form a very distinctive part of the topography. A cross-section of a typical valley in a glaciated region is shown in Fig. 8.1. It should be noted that the vertical dimension of the cross-section is greatly exaggerated.

Drainage design factors

All land drainage improvements need to be designed in such a way that they are compatible with the properties of flowing water. For larger improvement projects the nature and frequency of peak flows in the catchment must be evaluated with great care. In all projects, channel dimensions must be matched to likely discharge values if they are to remain satisfactory. Rapid flow will erode the channel and insufficient flow velocity or a marked reduction in velocity along a channel will result in deposition and blockage.

Drainage in river flood plains

Flood plains are complex parts of the landscape which can pose special problems for land drainers. These difficulties include lack of surface gradient and suitable outlet for drainage water; frequent flooding which may require flood protection banks; shifting river channels; return flows from the main channel through underlying gravel layers which cancel the benefits of land drainage installations; various springs and other inflows in the valley bottom or along marginal bluffs; and also the difficult nature of the soil profile which may be composed of silt with layers or lenses of peat.

Figure 8.1   Cross section of river flood plain

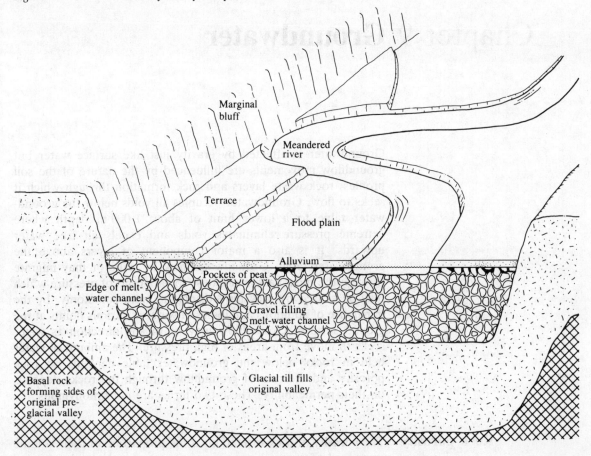

Chapter 9 **Groundwater**

Groundwater is affected by gravity just like surface water but groundflow movements are influenced by the nature of the soil profiles, rock waste layers and rock formation through which it seeks to flow. Groundwater occupies all voids below the groundwater table to a lower limit of about 1 000 m depth where extreme pressure eliminates voids and expels all free water upwards. It is also a major component of layers above the groundwater table. Groundwater may be static or may migrate through the rock mass in response to variations of pressure which may be caused by rock movements or, more commonly, by the imbalance of additions and losses of water at various points at the ground surface. Groundflow may be in any direction as water seeks to move from areas of high pressure. The amount of water held in rocks below the groundwater table depends on rock porosity and the rate of groundflow through the rocks depends on their permeability. These two terms are not synonymous.

Rock porosity

The **porosity** of rock formations is the proportion of their volume composed of spaces between constituent particles with its value only partly influenced by particle size. Uncompacted sedimentary formations of sand, silt, chalk and clay have up to 40 per cent pore space depending on packing arrangement, shape and uniformity of particle size. Compaction greatly reduces pore spaces. Sandstones can have up to 15 per cent of pore space, siltstones and shales about 5 per cent and slates rarely have more than 3 per cent. Many rocks have lost all porosity because secondary materials, like calcium carbonate, carried in by groundwater, have filled the voids. Similarly, the crystalline igneous and metamorphic rocks are usually non-porous. Clay usually has a higher value of porosity than sand and in a saturated condition will hold a greater volume of groundwater.

Rock permeability

The **permeability** of a rock formation is a measure of the energy required to cause liquids to flow through its bulk. In homogeneous formations the permeability may be the same in all directions but in layered rocks the permeability usually is greater along a bedding plane than across the layers. Just as described for soil profiles, the permeability of rocks depends on the presence of macropores, either of textural or structural origin. Where macropores are absent, groundflow movements, if they occur at all, are by capillary diffusion only. For a given amount of energy the distance travelled by groundwater through uncompacted sediments ranges from tens of metres per day in coarse sands to zero in clay. Also, as in soil profiles, compaction reduces permeability in rocks while structures – mainly jointing – greatly increase it. Sandstones have a permeability of up to 10 m/day while siltstones, mudstones, and shales have little permeability. Any rock formation with open joints can be very permeable.

Aquifers and aquicludes

Rock formations which can be used as a source of water are called **aquifers**. These have to be rock formations which allow appreciable flow movements towards a well and, therefore, have interconnecting macropores. In aquifers the groundwater is strongly influenced by gravity and flow movements are relatively rapid. Rock formations that lack macropores restrict water movements. Some may contain appreciable quantities of water but it is so firmly held that gravity flow movements hardly occur. The water, effectively, is a permanent part of the rock structure and no supply of water is obtained. Such rock formations are called **aquicludes**. Only aquifers can have a groundwater table. All types of impermeable rocks act as barriers to groundflow. These commonly include clay formations, siltstones, mudstones and non-jointed crystalline rocks. Drifts like glacial till and solifluxion layers have similar properties.

The properties of aquifers
Zones of saturation

Where an aquifer of, say, unconsolidated coarse sand occupies the ground surface on a level site such that it forms a self-contained system with no gains or losses of groundwater at the margins of the area being considered, the groundwater table will rise and fall in response to changes of weather. Periods of rainfall will cause the groundwater table to rise towards the surface and periods of drought will cause it to sink lower in the aquifer. In such a self-contained system, losses of groundwater from depth can result only from transpiration so that, below the depth of root penetration, there can be no loss of water. In this situation the groundwater table must lie somewhere within the rooting zone of plants. This kind of self-contained surface aquifer is uncommon since gains and losses from the margin of the area nearly always occur and have an influence on the location of the groundwater table. However, where such conditions do exist, provided that rainfall occurs evenly over the whole surface, the groundwater table will remain approximately horizontal at all times. As the groundwater table in any surface aquifer moves upwards and downwards in response to seasonal and shorter term changes of weather the lowest level reached in the aquifer marks the upper limit of the zone of **permanent saturation**. The highest level of the groundwater table forms the lower limit of the zone of **permanent non-saturation** while between these limits lies the zone of **intermittent saturation**. Any hollow which takes the ground surface below the level of the zone of permanent non-saturation will have surface water when the groundwater table is above the ground surface. If the hollow is lower than the upper limit of permanent saturation there will be permanent standing water.

Groundwater table adjustments

The groundwater table remains horizontal in this closed system only as long as the whole surface receives equal amounts of rainfall. Should a local storm affect only a part of the area being considered, the groundwater table in that locality becomes elevated above the general level. For a time this creates a hydraulic head within the aquifer and the hydraulic gradient initiates lateral groundflow. Groundwater flows through an aquifer according to **Darcy's Law**, the simplest form of which states that the rate of flow through a given volume of rock is the permeability value multiplied by the hydraulic gradient, expressed in convenient terms like litres per cubic metre of rock per day. The flow rate obtained from any rock may be the result of a large volume of water travelling slowly or a small volume moving quickly. The velocity of flow may be obtained by dividing the flow rate by the porosity value.

Head loss

In the highly permeable aquifer being considered the fairly rapid groundflow will quickly restore equilibrium and will dissipate any hydraulic gradient that may form. But flow will cease at some point before a true level is reached because some energy is lost in overcoming the small flow resistance in the aquifer. This factor is called the **head loss** which, for a given value of rock permeability, is directly proportional to the length of journey that the groundwater must travel. The head loss is inversely proportional to the permeability since the greater friction absorbs more of the potential energy available in the initial hydraulic gradient. Head loss affects the shape of the groundwater table. Generally, any irregularities that may occur in the groundwater table due to the uneven distribution of rainfall are lost in time as showers cover the whole area.

Topographical effects

In this highly permeable surface aquifer any head loss will always be insignificant and the groundwater table will remain more or less horizontal irrespective of the surface topography. But in less permeable aquifers the groundwater table will be horizontal only where the surface is reasonably level. Where there are surface undulations, additions of rainfall to the groundwater below the elevated parts of the landscape will exceed the rate of lateral flow, so that the groundwater table rises through the profile until the increasing hydraulic gradient matches the resistance to flow. This ensures that the groundwater table is influenced by the surface topography and takes the form of a more or less flattened image of the shape of the ground surface. The lower the permeability of the aquifer the more closely the groundwater table follows the shape of the ground surface. Since hydraulic gradient is always tending to cause groundflow from more elevated parts of the landscape towards lower parts, these lower parts tend to gain groundwater and the zone of intermittent saturation is narrower. In hollows the groundwater table tends to be less variable than in more elevated sites.

Lateral gains and losses from an aquifer

A highly permeable aquifer forming the ground surface usually is not a closed system because groundwater may be lost to or gained from nearby areas of different character and any such gains or losses will affect the depth of the groundwater table. The presence of a natural drainage channel along one margin of the area, deep enough to penetrate the zone of permanent saturation, will allow constant escape of groundwater. As long as the groundwater in the aquifer is sufficiently re-charged with rainfall the hydraulic gradient will maintain an outwards flow into the channel and the groundwater table will slope downwards in a gentle curve to the level of water in the channel. The gradient of slope of the groundwater table is inversely proportional to the permeability of the aquifer so that the presence of a drainage channel in an aquifer influences the depth of the groundwater table at a greater distance from the channel as the permeability increases. This is an important concept for land drainage design and is illustrated in Fig. 9.1.

Figure 9.1   Slope of groundwater table near an escape channel

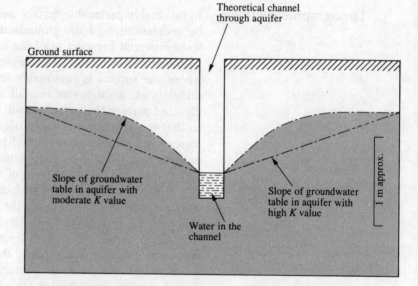

In a similar manner, a well sunk into a surface aquifer to supply water will form a **cone of depression** in the groundwater table; the rate of extraction must not exceed the rate of flow through the aquifer if a steady supply is required. Temporary loss of supply occurs when the cone of depression coincides with the base of the well. Conversely, where an aquifer receives groundwater from a nearby area the inflow will raise the groundwater table along that margin. Any surface aquifer receiving inflows of groundwater at one margin of its area and losing groundwater along the opposite margin will have a persistently sloping groundwater table and a constant through-flow of groundwater.

Figure 9.2 Ground flow towards drainage channels

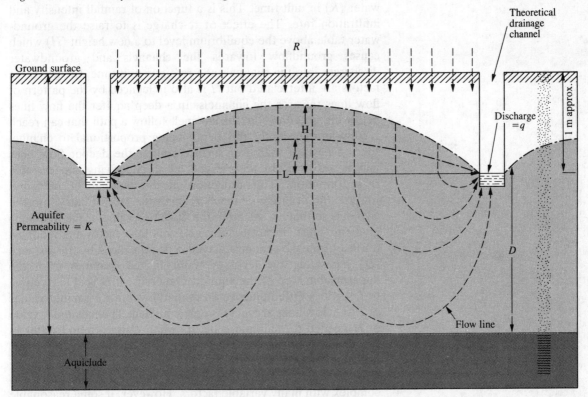

Groundwater movements in response to drainage

Figure 9.2 shows a relatively deep surface aquifer saturated to near surface level. When drainage channels are cut across the aquifer, deep enough to penetrate the zone of saturation and close enough to influence the groundwater table over the whole area, groundflow occurs towards the channels along pathways or **flow lines** as shown in the figure, which is a cross-section of the profile set at right angles to the line of channels. Flow continues until the groundwater table is lowered to the point of equilibrium at a height (*h*) above the level of water in the channels mid-way between channels. The value of *h* is influenced by the permeability of the aquifer (*K*), by the distance between channels (L) and by resistance to flow into the channels caused by the concentration of flow lines towards the channel. No further flow can occur until the aquifer receives more rainfall. When significant rainfall occurs the aquifer is re-charged by an amount of ground-water (R) in unit time. This is a function of rainfall intensity and infiltration rate. The effect of re-charge is to raise the ground-water table above the equilibrium level to a new height (H) which causes groundflow towards the channels and groundwater discharges into the channels at a rate q in unit time. Clearly these factors are interrelated but H is also influenced by the pattern of flow lines towards the channels. In a deep aquifer the flow lines below the water level in the channels follow a path that can reach a considerable depth, this depth being proportional to channel spacing L. In a shallower surface aquifer the deeper flow lines may be restricted by the presence of an underlying impermeable rock formation. This reduction in the number of pathways towards the drainage channels reduces the rate of entry into the channels so that, to maintain the discharge value q, the hydraulic gradient must increase. Because of this factor, groundflow towards drainage channels may also be influenced by the distance (D) separating the level of water in the channels from the impermeable base of the aquifer. Generally, if D is at least equal to $L/4$ it has little influence on groundflow but for smaller values of D the flow lines are increasingly inhibited. The maximum value of H needed to maintain q occurs when D is zero and water in the channels is at the same level as the impermeable layer. Attempts to provide mathematical solutions for assessing ground-flow towards drainage channels have shown that the process is complex with many variable factors. However, if some reasonable assumptions are made, a less complicated formula can be used and a satisfactory method of calculating drain spacings (L) can be obtained. Selection of drain spacings is discussed in Chapter 15.

Surface aquicludes   Where the formation occupying the surface is an aquiclude there is no groundwater table and the groundflow patterns discussed above for aquifers do not occur. In most cases only the A horizon is permeable and any gravitational water in that horizon is trapped forming a perched watertable. When drainage channels are cut through a surface aquiclude, gravitational water in the A horizon may flow horizontally along the base of the permeable layer until it reaches the channel. The profile acts as a perched watertable whether the horizons below the permeable layer are saturated or dry. There is no gravitational water to cause ground-flow. Evaporation and capillary seepage may cause the side walls of the drainage channel to dry out and cracks will appear to provide pathways for groundflow. Similarly, fractures in the B horizon caused by soil structures may deepen the permeable zone in the profile. However, soil fracturing cannot change a surface aquiclude into an aquifer unless they penetrate to the level of water in the drainage channels in which case the profile behaves as a shallow surface aquifer with the impermeable base at the level of water in the channels (when D=0). This is a very unusual situation in practice. Soil fracturing rarely reaches the depth of drainage channels.

Groundwater pressure

Returning to a uniform surface aquifer which has not been drained, the groundwater table is free to rise and fall unhindered and water pressure along the upper limit of the saturated zone, at the groundwater table, is at atmospheric pressure (in relative terms, that is, at zero pressure). All water at a lower level is subject to a positive pressure which is proportional to its depth below the groundwater table while water in the profiles above the groundwater table is subject to a negative pressure or is held at tension. The degree of tension is proportional to its height above the groundwater table. The practical implication of this is that highly permeable soils may be over-drained if the drainage channels are too deep. The depth of film of available water is reduced. Groundwater tables which are free to rise and fall according to hydraulic gradients (i.e. to variations in water pressure) are found in surface aquifers which may be described as **unconfined aquifers**. The groundwater table is at zero pressure everywhere and is said to lie at its **phreatic surface.**

Many aquifers occupy positions below the surface and are overlain by other rock formations. Inflows of groundwater may cause the groundwater table to rise to the upper limit of the aquifer and may seek to rise into the formation above. If the overlying formation is impermeable the groundwater movement will be restricted and a **confined groundwater table** will exert pressure on the formation above. This type of rock structure is called a **confined aquifer** and a vertical pipe set to penetrate the confining layer will allow groundwater to rise up the pipe a distance above the aquifer that is proportional to the pressure of the confined groundwater. The water rises to reach the **piezometric surface** of the groundwater as shown in Fig. 9.4. If the confining formation is not totally impermeable it will form a **semi-confining** layer (actually it may be quite permeable as long as it is markedly less permeable than the particular aquifer being considered). The groundwater is held below its piezometric surface but it may rise some way into the semi-confining formation. This type of structure is called a **semi-confined aquifer**.

The effect of rock formations below an aquifer

The effect of an impermeable barrier below an aquifer has been considered only in terms of its effect on groundflow towards a drainage channel. In horizontal rock formations any lateral component of groundflow is in response to a hydraulic gradient. Truly horizontal sedimentary successions are relatively rare but include recently exposed marine sediments like those of the much studied Polders in the Netherlands. More common are sedimentary successions tilted during the process of uplift and outcrop at an angle to the surface. Each formation dips away from the surface and the aquifers fill with groundwater to a level that is determined by the elevation of the lowest escape pathway for groundwater. Each aquifer has its own level of groundwater table and is not influenced by the level of water in nearby aquifers if they are separated by aquicludes. The presence of an inclined impermeable layer below an aquifer is important where the sloping surface provides a pathway towards an escape route for groundwater. This is seen in Fig. 9.3. In this type of structure the groundwater table in the aquifer is held at a very low level because the gradient of the impermeable surface causes lateral flow.

Dipping aquifers as a source of water

Dipping aquifers can be good sources of water when tapped by wells of sufficient depth. This is illustrated in Fig. 9.4. A well sunk into the outcrop area of the aquifer may reach the groundwater table with little effort and obtain a supply of water. Away from the outcrop area the well may have to be deeper but water will rise in the well to reach the piezometric surface giving a better supply. As discussed in Chapter 6, outcropping aquifers often form more elevated parts of the landscape. In some cases the groundwater table in the elevated aquifer may be higher than

Figure 9.3   *Ground flow in a sloping aquifer*

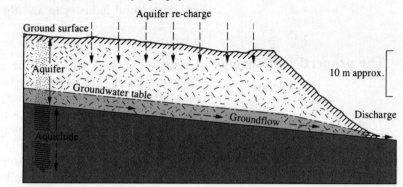

Figure 9.4   *Wells in a dipping aquifer*

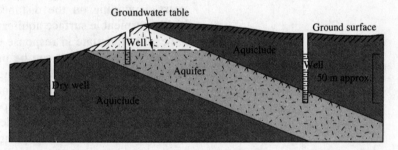

the ground surface in nearby lowlands so that the piezometric surface may be above the ground surface. A well sunk into the aquifer will fill to ground surface and overflow. This is called an **artesian well**. All of these features are shown in Fig. 9.4.

A major artesian water supply is provided by permeable rocks that form the Blackhills in Dakota. The aquifer dips below the plain to the east and provides water for a large area. Where an aquifer is part of a syncline – a downfold in the rocks – there may be an elevated area of recharge at each end of the formation. This kind of structure is called an **artesian basin** and provides a good water supply to the valley area between the outcrops of the aquifer. The chalk formation in south-east England at one time supplied artesian water to the London area but over-exploitation has lowered the piezometric surface below the valley ground level and pumping is now necessary. A very large artesian basin occurs in Queensland and nearby areas in Australia and is the basis of an agricultural industry in an otherwise dry region.

The properties of springs
Occurrence

Groundwater may return to the surface naturally at a range of sites and the spring issues may take the form of patches of wet land, distinct upwellings of water which spread over the surface or larger flows which are the source of a stream. In limestone country major streams may emerge from caverns. Springs may occur randomly at the surface but seepages marking the lower margin of an outcropping aquifer may occur in a distinct zone across the landscape which is called a **springline**. The interval of time required for water to move from the area where it percolated into the rocks to the area where it finally re-emerges to the surface can be anything from a few hours to tens of thousands of years depending on the distance and rate of flow. Only small, highly permeable surface aquifers with short travel times can have issues that fluctuate in response to short-term changes in climate. Larger aquifer systems can be influenced by seasonal changes of climate but usually after a time lapse. Change of level of the groundwater table is important in some systems. **Permanent springs** are fed by the zone of permanent saturation while **intermittent springs** (sometimes called bournes) are fed by the zone of intermittent saturation. Springs fed by water that has travelled a considerable distance are little affected by seasonal climatic change. Any stream or drainage channel that tends to run water constantly, irrespective of soil water status, is likely to be fed by groundwater.

Classification

For land drainage purposes springs may be grouped into two basic types, those that result from gravitational groundwater movements and those caused by artesian pressure in a confined or semi-confined aquifer. These are called **gravity springs** and **artesian springs** respectively. Gravity springs, in turn, can be divided into two types according to the rock structures causing the seepage. Some springs result from the presence of **easy pathway structures** which provide an escape route for groundwater. In other situations an obstruction in an aquifer may cause a build-up of groundwater causing it to reach the surface near an **overtopped barrier structure**.

Easy pathway springs A number of rock structures and related topographical features provide an easy pathway for the escape of groundwater to the surface.

(a) Where an aquifer occupies the surface, groundwater will escape wherever the groundwater table intersects the surface. If the escaping water is retained in a hollow the flow will cease as soon as the hollow fills to the level of the groundwater table. If, however, the water on the surface is not retained at the site of escape, seepage will continue as long as the groundwater table is high enough. Any site with sloping ground, a natural channel or an excavation that takes the ground surface below the groundwater table allows groundwater seepage.

(b) Sedimentary successions which are horizontal or dip towards the outcrop surface as shown in Fig. 9.3 provide escape routes for groundwater. Springlines can occur at the base of each outcropping aquifer.

(c) Igneous sills contained in a sedimentary succession can act as an aquifer or as an aquiclude depending on their permeability. Usually they are massive and springs may occur at the top margin of their outcrop. Some types of magma contract when they cool and split into columns causing the sill to be highly permeable and water may seep from the lower margin of the outcrop. Sills usually are more resistant to weathering than the nearby sedimentary rocks and tend to form prominent features in the landscape.

(d) Any kind of hard rocks – including igneous and metamorphic – occupying surface positions in the landscape are usually very permeable because of the network of fractures. They behave as surface aquifers. Groundwater can escape from any fracture that provides an easy pathway to the surface. A fault through the rock may collect groundwater from a large number of joints and can be the source of a stream.

(e) Limestone formations are an extreme type of rock rendered permeable by fractures. Groundflow through the fractures dissolves the rock to form channels and caverns which drain water towards the base of the outcrop where it may emerge as a stream or river.

(f) In some formerly glaciated lowlands a once extensive cover of glacial till may be eroded by more recent surface water flow leaving a series of isolated cappings on any more elevated part of the landscape. Groundwater often seeps from the lower margins of the till-covered areas because the lowest horizons of the till usually contain a high proportion of rock rubble making them permeable.

(g) Fluvioglacial mounds resting on glacial till are common sites for easy pathway springs. In most cases groundwater seeps from the lower margin of the outcrop and spreads over the surface of the glacial till. This kind of seepage is illustrated in Fig. 9.3.

Overtopped barrier springs

Various rock arrangements have a damming or blocking effect on groundflow through an aquifer which causes a build up of groundwater. In some situations the groundwater may rise to the surface as it overtops the obstruction. Springs of this type are often intermittent and flow only when the groundwater table is sufficiently high in the aquifer. The various blocking structures include:

(a) A sedimentary succession dipping away from the outcrop surface with no other escape routes. Each aquifer may fill to the level of the outcrop and overflow as a springline.

(b) Various kinds of barriers in the path of groundflow through a surface aquifer. A number of rock structures cause this type of spring to occur and include minor folds in a sedimentary succession which bring an impermeable formation nearer to the surface; a fault through a sedimentary succession carrying a block of impermeable rock nearer to the surface; a fault through an aquifer containing bands or lenses of clay which may bend to form vertical barriers instead of fracturing; or an igneous dyke cutting through a sedimentary succession.

(c) Fault lines in ancient rocks may be filled and blocked by cementing minerals converting the fracture into a barrier to groundflow.

All of these barriers act in the same way. Groundwater is dammed back behind the barrier and the groundwater table in the aquifer rises until the soil horizons are saturated. An example of an upfault block is shown in Fig. 9.5.

Artesian springs

Artesian springs all result from the same basic arrangement of rock formations illustrated in Fig. 9.4. The essential features are:

Figure 9.5   *Overtopped barrier spring*

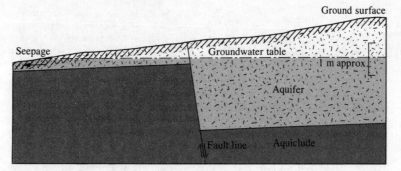

1 A confined or semi-confined aquifer.
2 An elevated exposed area of the aquifer which can be recharged with rainfall.
3 A rate of recharge which at least equals the total losses from the system.
4 Sufficient permeability or the presence of fractures in the confining layer to allow the upward escape of groundwater.

The confining layer need not be particularly impermeable and the aquifer need not be especially permeable to cause artesian seepage. All that is required is that the upper layer be less permeable than the main aquifer. Fine sand overlying medium sand, or clay overlying jointed shale, are equally effective artesian structures. The greatest upwards pressures, however, are likely to occur where a highly permeable aquifer is confined by a totally impermeable confining layer. In all situations where the confining layer is impermeable the groundwater can escape only at isolated points where the formation has fractures or is otherwise punctured. Artesian springs take the form of clearly defined upwellings in the middle of patches of boggy land. In some cases the site of upwelling is marked by a dome-shaped hummock composed of rock waste carried up by the water and called a **spring mound**. In semi-confined aquifers the leakage may be more diffuse and can be mistaken for a topographical high groundwater table. The pattern of flow lines of artesian seepage through a semi-confining aquifer are shown in Fig. 9.6.

Artesian structures in sedimentary rocks

Artesian springs, like gravity springs, may be caused by various rock structures. Within sedimentary successions, confined

Figure 9.6   Flow lines from a semi-confined aquifer

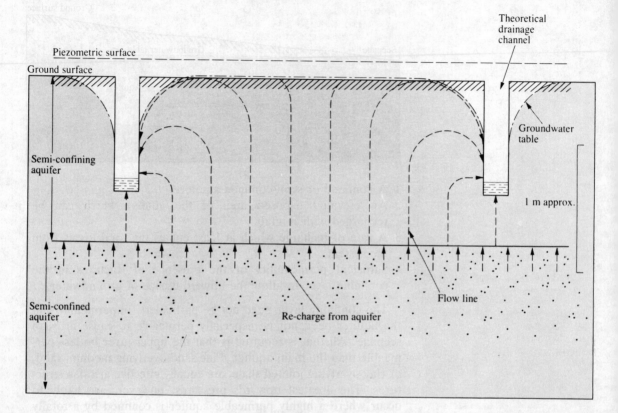

aquifers usually do not leak because the confining layers above form a complete seal. Exceptions may occur where erosion has made a confining formation very thin. This is seen in some clay vales where further removal of cover, for example by excavating a major drainage channel, may be sufficient to puncture the sealing layer or the pressure of groundwater may cause distortion of the drainage channel. Where a sedimentary succession is composed of thin formations artesian seepage is more likely to occur. The type of rock structures most likely to be involved are thin marine sedimentary layers, alluvial fans, river delta deposits and river flood plain deposits.

Artesian structures in rocks covered by drift

Wherever the land surface is covered by a drift formation of limited permeability any spring seepages that may have been caused by the underlying rock formations are sealed off. The drift acts as a semi-confining or confining layer and where it overlies any kind of aquifer the basic requirements of an artesian structure may be present. Drift formations of limited permeability include loess, solifluxion deposits and glacial till. All types may cause artesian springs but the largest numbers are found associated with glacial till-covered maritime regions. Springs found in such areas are not all artesian seepages but they form a gradational series and it is convenient to consider them as a special group.

Springs associated with glacial drift

Figure 9.7 is a cross-section of a valley or the margin of a plain in a maritime region that has been affected by glacial erosion and deposition.

Springs may occur at a number of points on the landscape

● Groundwater may escape from any easy pathway structure in the upland area above the level of the glacial till forming an easy pathway spring.

Figure 9.7   Spring structures in a glacial drift

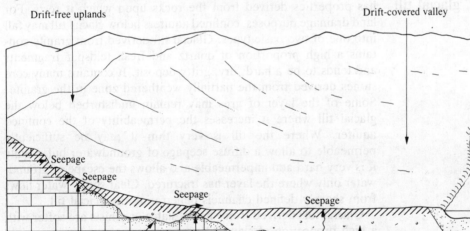

- Where the rock forming the upland drift-free area is uniformly permeable, the most common escape route for groundwater is some low point along the upper margin of the glacial till cover when the groundwater table in the aquifer is high enough to form an overtopped barrier spring.
- The surface upon which the glacial till rests is not always even. All manner of ribs and projections of hard rock may be present, over which the smoothed out glacial till is necessarily much thinner. Near the upper limits of the drift-covered area where the till is thinning out towards its margin, the more prominent rock projections may create gaps in the till through which groundwater can escape. Their presence high up in the valley sides may control the level of the groundwater table in the aquifer in which case overtopped barrier springs occur.
- At all other sites below the level of the piezometric surface in the drift-covered area, any thin areas or any fractures in the drift may be the locations of artesian springs.

Artesian structures in glacial till

Glacial till usually has not been moved far by the ice and often has properties derived from the rocks upon which it rests. For land drainage purposes, confined aquifers below glacial till may fall into one of two basic types. Glacial till derived from granite contains a high proportion of quartz and fresh feldspar fragments and tends to be a hard, dry, gritty deposit. It contains many core stones derived from the partially weathered zone in the granite. Some of the layer of grus may remain undisturbed below the glacial till where it increases the permeability of the confined aquifer. Where the till is very thin it may be sufficiently permeable to allow a diffuse seepage of groundwater but usually it is very hard and impermeable and allows the escape of groundwater only where the layer has fractured. Clear spring water flows from sharply defined channels penetrating the glacial till.

Glacial tills derived from shales, slates or schists are softer with a high proportion of clay and/or mica minerals and contain only small fragments of rock. The underlying rocks usually are not weathered but loose fragments have collected to form a highly permeable layer below the glacial till. This type of glacial till is a clay formation that is relatively unstable and where artesian seepage occurs it often takes the form of an area of slurried clay which slowly oozes upwards to form a spring mound at the surface.

Other hard rocks can form confined aquifers resembling one

or other of these two types depending on the nature of the minerals present. On the other hand, sedimentary clay formations are not permeable and when overlain by glacial till derived from the clay formation it is difficult to distinguish drift layer from sedimentary formation. No confined aquifer is formed and artesian seepage is unlikely. In any type of rock an artesian seepage is unlikely where the cover of glacial till is very deep although upwellings have been observed through layers at least 6 m deep. Spring issues through the deeper confining layers may be associated with the powerful pressures occurring where highly permeable aquifers connect to a recharge area at considerable height.

Identifying spring seepages
General features

Drainage work to control groundwater movements needs to be matched to the type of rock structures present if it is to be effective. For this purpose some knowledge of local geology is needed. This requires consideration of three basic factors.

1 Recognition of any drift formation and assessment of the depth and permeability of the drift layer(s).
2 Identification of the underlying rocks.
3 Identification, as far as possible, of the relevant rock structures.

Most of the information needed concerning the rock formations may be obtained from geological maps which identify the outcropping rocks and important features like fault lines and the angle and direction of dip. Most maps also identify any drift present.

Observation of the pattern of spring issues is important. In particular, distinct spring-lines associated with a change of soil type or a change of gradient will usually support the evidence of the map. Confirmation will be obtained by examining the profiles revealed by test excavations in the area. These should be sufficiently deep to reveal the nature of any aquifer supplying the groundwater where this is practicable. Close attention to changes of profile along the open trench of the drainage channel, as it is excavated towards the wet area, will provide much valuable information.

Gravity springs

Gravity springs are often associated with changes of soil type and changes of gradient. This is particularly common in all types of spring seepage produced by sedimentary successions since the rocks that cause the springs also influence the topography and provide the rock waste in which soils are formed. It is less evident in other types of rocks. In all cases test holes will reveal the structure causing the springs. Excavations above the spring seepage will tend to reveal the symptoms of a high groundwater table in a permeable subsoil while, just downslope from the spring, a saturated profile will have all the symptoms of a perched water-table over an impermeable subsoil or over some type of impermeable rock formation not far below the surface.

Artesian springs

Artesian springs are not usually associated with changes of soil type or of gradient but tend to occur in random fashion across a drift-covered surface. The rock structure may be confirmed by digging test holes in the vicinity of the springs. When the excavations reach and expose the confined aquifer, water flows into the hole, sometimes vigorously, but in other cases gently filling and overflowing from the hole. Where the ground surface lies near to the piezometric surface the upwelling may not reach the surface of the hole and the spring may be intermittent as the piezometric surface rises and falls with the seasons.

Drainage works to control groundwater

Drainage design must seek to match precisely the nature of the groundwater movements, causing a land-drainage problem. A high groundwater table in a surface aquifer requires a regular system of parallel drainage channels set close enough to control the zone of saturation over the whole site. Control of lateral groundflow through an aquifer requires an intercepter channel near to the receiving margin. Where the supplying area is more elevated than the land to be drained or where the source is a high-level body of water nearby (e.g. a canal) it is necessary to take account of deep flow lines which may pass below the interceptor channel. Careful selection of the line of the channel is required. Where springs cause a drainage problem the drainage channel must be set to intercept groundflow within the aquifer so that groundwater cannot reach the soil horizons. Drainage of artesian seepage may require relatively deep channels. Drainage channels set at regular spacings across the field are needed to control a high groundwater table or a perched watertable but such installations need not be suitable for the control of lateral groundflow or upwards seepage as shown in Fig. 9.6. Drainage to control groundflow movements usually requires special techniques.

Part two **Practical land drainage**

Chapter 10 **Ditches**

Open ditches form at least a part of most land-drainage works. The name **ditch** in a restricted, land drainage context refers to a man-made drainage channel that is open to the surface and is used to collect and carry water drained from the land. Larger channels constructed mainly to form a means of transport are called **canals** but the two definitions are often merged and terms like drainage canal are seen sometimes in land drainage literature. For drainage purposes the design requirements are identical and no confusion need arise.

Advantages and disadvantages

Ditches have been used by land improvers over many centuries for a variety of purposes. They have several advantages over other forms of land drainage aids. Ditches allow easy entry of any surface water into the drainage channel, they have an overload capacity to cope with storm conditions and an open channel allows easy access to any associated pipe-drainage system. However, ditches permanently occupy land which might be valuable, they restrict field traffic and need careful and regular maintenance to remain effective. As a general rule each farm should have the minimum length of ditches that can efficiently permit removal of excess soil water.

Functions

The ditches found on most farms are intended to carry out one or more of the following functions:

- To act directly as land drains.
- To control the groundwater table either alone or associated with a pipe-drainage system.
- To act as collectors of water removed from the soil by pipe-drainage systems.

- To intercept surface flow or groundflow.
- To carry water from one area to another for drainage or irrigation purposes.
- To deliver irrigation water on to the land.
- To act as reservoirs to hold water for irrigation or land-drainage pumps.
- To act as soakaway pits.
- To provide water for livestock.
- To act as stock barriers or to mark the boundaries of farms.
- To provide soil for earth banks and hedgerows along field boundaries.

Land drainage ditches

A regular system of open channels could be used to drain land quite successfully as **land drainage ditches**. The ditches would need to be carefully graded from the outlet point, cut deep enough to cope with the particular drainage problem and close enough to drain the between-channel areas as determined by the soil properties. This, of course, breaks up the land into very small parcels and would be unsuitable for most farms. A type of land drainage based on this method is used for poor quality grazings where a regular system of open channels is cut by means of a heavy duty plough. This method is discussed in Chapter 13.

An early form of land drainage by closely spaced channels was achieved by repeated ploughings to throw the land surface into a series of parallel ridges and furrows with ridges 5–10 m apart. Water is shed from the ridges and flows along the furrows. Ridge and furrow land can still be seen in old farming areas where the uneven surface is a nuisance to a land improver in modern conditions. A modern version of ridge and furrows drainage is to grade the land surface into a series of wide strips so that all surfaces shed excess rainfall into a system of parallel ditches.

Groundwater table control ditches

Land with a high groundwater table is usually flat, often over a wide area, with very little gradient available to carry away drainage water. Such land may require **groundwater table control ditches** because the open channels can remain effective even with very slight gradients. As long as there is an outfall point, the ditch gradients are even and the channels are kept clear, the water will flow away. In the most permeable soils well-constructed ditches may control the groundwater table unassisted but usually they will be associated with a pipe-drainage system.

Collector ditches

The most common use of open channels under modern conditions is that of a **collector ditch** to receive water from the field drains and to remove excess water from the site.

Interceptor ditches

Surface flow and interflow water movements can be controlled by well-placed **interceptor ditches**. Open channels excavated across the general slope of the land, along the upper margin of the area to be protected, easily divert all flows at or near the surface. Interceptor ditches are of less value for most groundwater movements because the groundflow tends to carry soil particles into the channel or otherwise distort its shape and gradient.

Other ditch functions

Farm ditches deliver drainage water into natural water courses or into larger ditches constructed to serve the needs of several farms. Generally such **carrier ditches** should not be altered unless the work is controlled by an engineer. Carrier ditches include channels used to deliver irrigation water to the area required. In the more mechanised agricultural areas irrigation water is usually applied to the land by mobile pipes and sprinklers so that small field delivery channels are not used. **Storage ditches** for irrigation or pump-drainage systems are an integral part of the soil water control installation and these also must be left intact. Most other types of open channels encountered are not primarily concerned with drainage and can be piped and/or filled when required and more satisfactory installations used to carry out their functions.

New ditches

The number of drainage projects requiring new ditches in the long-established farming areas is not great since nearly all schemes make use of existing channels. Where new channels are required for a single farm they will nearly always be relatively short collector or interceptor ditches. Larger projects involving several farms and requiring a new carrier ditch are rare and mostly they are part of some major project like the reinstatement of agricultural land following some industrial activity such as open-cast coal extraction. The great majority of drainage schemes involving work on ditches are concerned with the improvement of existing channels, very many of which are neglected and in poor condition.

Ditch design When new ditches are excavated or existing channels are improved, the new dimensions, as far as possible, should provide a channel that is most suited to the likely discharge value, otherwise the flowing water will tend to adjust the channel until these requirements are met. Those channels which are least in harmony with the discharge will be altered most rapidly and further work to correct the fault will soon be required. Good ditch design must also cater for the needs of all existing and future land drainage works associated with it. Consideration must be given to the type of material through which the channel is cut and the route chosen should be compatible with other aspects of good land management. Perfect channels are never achieved but careful planning can enable the land improver to come near the ideal solution.

Design factors for ditches The required dimensions for a new or reconditioned ditch are determined by the following factors.

- The run-off of the catchment area served by the ditch. This is a function of the chosen return frequency of rainfall events, the area of the catchment; the topography, the nature of the soil and underlying rocks and the land use.
- The channel discharge which is controlled by the run-off and channel gradient.
- The ditch bottom width which must be suitable for the discharge.
- The depth of the ditch which depends on local topography and land drainage requirements.
- The slope of the ditch banks which must remain stable; this depends on the type of material through which the ditch is cut.
- The top width of the ditch is then established automatically when all of the other requirements are met.

Ditch discharge For design purposes:

Discharge = Catchment run-off + Inwards groundflow + Other inflows

For small catchments of up to several fields size of agricultural land only:

Discharge = (Design rainfall rate × Area of ditch catchment) + Inwards groundflow + Other inflows

Groundflows entering the ditch directly are usually not significant and for most sites can be ignored. Where test excavations reveal significant inflows of groundwater it may be necessary to increase ditch capacity by an equivalent amount. It should be noted that groundwater may arrive from outwith the surface catchment area. In a similar way, any pipes carrying water from outwith the catchment that are to discharge into the channel must be considered.

For larger catchments where the time of concentration of peak flow rates is important or for any site where buildings, public facilities or matters of safety are involved, the channels must comply with the standards specified by qualified engineers.

Ditch bottom width

Bottom width must match the discharge and there should be a suitable ratio between depth of flow and width of flow. A reasonable choice of bottom width would provide a flow width about three times flow depth at peak design flow rates and more than three times flow depth at smaller discharge rates. If the bottom of the ditch is cut too wide the flow will occupy only a part of the channel and will erode a new, smaller channel in the bed leaving the rest of the bed to grow weeds. If, on the other hand, the bottom is too narrow, velocity of flow is increased and the discharge will tend to impose a better depth:width ratio by eroding the channel sides. Minimum discharge rates must also be considered. For many small catchments the base flow rate may be nil and flow will cease in a dry season. Where a channel is likely to be dry for long periods each year it may be better to use a pipe channel to avoid the cost of weed clearance.

Ditch bottom width may be calculated as follows.

- Choose the most appropriate design rainfall rate.
- Calculate the ditch discharge resulting from this amount of rainfall:

$$Q = 0.116 \times R \times A$$

where Q = ditch discharge in litres/sec
0.116 is a conversation factor to change mm/ha/day into litres/sec
A = area of catchment in hectares
R = design rainfall rate in mm/day

- Add the measured discharge of pipes delivering water from outside the area (litres/sec).
- Add the likely volume of any inflows of groundwater (litres/sec).
- Measure the gradient of the ditch line. For long ditches the gradient of the last section near its outlet (about one-third of ditch length) is more important than the average gradient for the whole length.

Using these values of design discharge rate and effective gradient choose a suitable bottom width from Table 10.1.

Take the example of an area of land one kilometre square and a design rainfall rate of 10 mm/day. This provides 10 000 m³ of water in 24 hours or 116 litres/sec. If the effective gradient is 1 : 500 the table shows that a bottom width of 40 cm is adequate. This will produce a flow with satisfactory dimensions. The overload capacity of an open channel can be illustrated by calculating the discharge from such a ditch 1 m deep running bank full. This could have a discharge of about 1.0 cumec before overtopping and would remove about 90 mm of rainfall from 1 km² of land

Table 10.1 Ditch discharge capacity in litres/sec where flow width is about 3× flow depth and the channel is in moderate condition

Ditch bottom width (m)	Gradient 1:												
	10 000	1 500	1 000	800	600	400	200	100	90	80	70	60	50	40
0.2	6	16	18	21	24	29	41	59	62	66	70	76	83	93
0.3	17	46	55	62	72	87	123	174	182	195	208	225	246	275
0.4	37	99	118	135	153	186	264	373	392	417	446	482	528	590
0.5	68	181	216	246	282	342	483	683	717	764	817	883	966	1 080
0.6	110	292	349	398	455	552	781	1 104	1 158	1 234	1 320	1 427	1 561	1 746
0.7	167	441	527	601	686	833	1 178	1 666	1 747	1 863	1 992	2 153	2 356	2 634
0.8	239	632	756	861	985	1 195	1 810	2 389	2 506	2 671	2 854	3 088	3 379	3 778
0.9	324	857	1 025	1 168	1 336	1 620	2 291	3 240	3 398	3 622	3 874	4 187	4 582	5 123
1.0	433	1 146	1 370	1 562	1 787	2 167	3 064	4 333	4 545	4 845	5 181	5 600	6 128	6 852

Note: Not strictly accurate for ditches with vertical sides or for ditches which have a batter greater than 1 in 3. The margin of error increases with discharge and for carrier ditches – an engineer should be consulted.

in 24 hours. In practice the limiting factor for the ditch would be the capacity of any culvert or other permanent constriction along its route. Generally it is not worthwhile to excavate a ditch requiring a bottom width less than 30 cm and situations requiring very small channels are more suitable for piped outlet channels, unless a small interceptor ditch is needed.

Ditch depth

The depth of a ditch is determined by the available outlet receiving the water carried by the ditch, by the topography of the site and by the design requirements of the field drains served by the ditch. The field drainage requirements should be considered first since they determine the success or failure of the whole project. The field drains should be at their design depth for the whole area to be improved and should have sufficient gradient throughout their length (see Ch. 11). The field drain outlets should be above the water in the ditch at all times except when the design return frequency has been exceeded by a considerable margin. This can be achieved by setting the drain outlets a minimum of 15 cm above the ditch water level at design discharge value as shown in Fig. 10.1 which shows also the main dimensions of a ditch.

On level sites the design requirements of field drains will influence the spacing between ditches as well as the depth of the open channels. All areas of slight relief and poor gradients will present difficulty and the design depth of the ditch may be achieved only at the expense of gradient or by installing drainage pumps. In any event it is better to reduce the gradient of the ditch than to place

Figure 10.1   Cross-section of a ditch

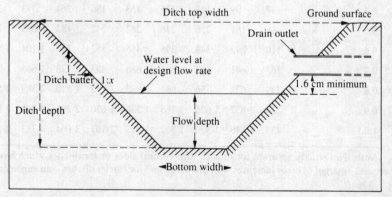

limitations on the pipe drains in the fields. On sites with a natural gradient there will be no limitation on achieving a suitable depth.

In general a ditch should not be less than 1 m deep to allow sufficient soil cover for the associated field drains. The exceptions are small interceptor ditches which have small discharges but, even with these, the depth should not be less than 75 cm.

There are many sites where a ditch must be much deeper, particularly where surface undulations must be crossed, and this will be acceptable to a maximum depth of about 5 m as long as other design factors are satisfactory. Deep ditches may have unstable banks, they can be dangerous and the channel occupies an excessive area of land. They can be avoided often by careful choice of route or by piping the channel when only short sections are involved.

Ditch gradient

The gradient is influenced by the nature of the site but in some cases a designer can choose a suitable route to obtain the best gradient. Many ditches are located on very flat sites where gradients must be constructed, but, provided the work is accurate, the channel can function with gradients as little as 0.01 per cent. Ideally, ditches should have sufficient gradient to cause flow velocities that give a self-cleansing effect without eroding the channel. Flow velocities of about 0.5 m/sec or less result in **siltation**, which is a build up of soil fragments in a channel, and rapid weed growth. Flows of 0.75 m/sec or more can prevent weeds from becoming established, while flows of 1.5 m/sec or more can erode the channels. Flow velocities produced by ditches are shown in Table 10.2. These are approximate values based on the assumptions that the flow is at design discharge rate and that the channel is smooth and clean. Velocities will be less when a ditch is in poorer condition.

Flow velocities at which erosion occurs depend also on the nature of material forming the channel. Resistance to erosion depends on soil particle size and cohesion between particles. Coarse sand, and clay particles are most stable while silt, very fine sand, fine sand and soft peat are more easily eroded.

Channels cut into the least stable materials can be eroded at flow velocities of less than 1.5 m/sec. In the worst cases it is difficult to keep the channel open because of bank collapse. In most farm ditches the flow velocity is insufficient to clear the channel so that regular cleaning is necessary. If the effect of

Table 10.2   Ditch flow velocities in metres/second where the channel is in a good, newly-cleaned condition

Ditch bottom width (m)	Gradients 1:													
	10 000	1 500	1 000	800	600	400	200	100	90	80	70	60	50	40
0.2	0.10	0.26	0.31	0.35	0.40	0.49	0.69	0.98	1.03	1.09	1.17	1.26	1.38	1.55
0.3	0.13	0.34	0.41	0.47	0.53	0.64	0.91	1.29	1.35	1.44	1.54	1.67	1.82	2.04
0.4	0.16	0.41	0.49	0.56	0.64	0.78	1.10	1.56	1.63	1.74	1.86	2.01	2.20	2.46
0.5	0.18	0.48	0.58	0.66	0.75	0.91	1.29	1.82	1.91	2.04	2.18	2.36	2.58	2.88
0.6	0.20	0.54	0.65	0.74	0.84	1.02	1.45	2.04	2.14	2.29	2.45	2.64	2.89	3.23
0.7	0.23	0.60	0.72	0.82	0.94	1.13	1.60	2.27	2.38	2.54	2.71	2.93	3.21	3.58
0.8	0.25	0.66	0.79	0.90	1.03	1.24	1.76	2.49	2.61	2.78	2.98	3.22	3.52	3.94
0.9	0.27	0.71	0.84	0.96	1.10	1.33	1.89	2.67	2.80	2.98	3.19	3.45	3.77	4.22
1.0	0.29	0.76	0.91	1.04	1.19	1.44	2.04	2.89	3.03	3.23	3.46	3.73	4.09	4.57

siltation is to be minimised it is essential, at construction stage, to ensure that the gradient does not decrease significantly towards the outlet end. Where steep gradients cannot be avoided the flow velocity may be reduced slightly by cutting a wider channel but this has only a marginal effect and usually erosion can be prevented only by constructing **weirs**, **drop structures** or **chutes** with baffles to absorb erosion energy. Alternatively, the channel may be lined with erosion-resistant material like stone aggregate. In all cases where the channel is large or erosion may be serious it is advisable to consult a qualified engineer.

Bank slopes   The gradient of a ditch bank is called the **batter** and all banks should have sufficient batter to ensure that channel sides do not collapse. Insufficient batter may allow the bank to collapse into the channel and this is called a **bank slip**. Ditches should be excavated with a batter that is less steep than the natural angle of repose of the bank-forming material. In most cases it is found that satisfactory batters have gradients (vertical distance: horizontal distance) as follows.

Rock, dry fibrous peat	1 : 0	to 1 : 0.5
Clay, loess	1 : 0.5	to 1 : 1
Loam, clay loam, silty loam	1 : 1	to 1 : 1.5

Compacted sand or sandy loam	1 : 1.5	to 1 : 2
Loose sand or sandy loam	1 : 2	to 1 : 3
Loose fine sand, soft peat	1 : 3	to 1 : 4

If the ditch is to be more than 1 m deep it is safer to use the least batter for each group of bank-forming materials. Where very flat batters are not practical or where unstable soils like very fine sand are present it is necessary to stabilise the banks (see p 139). Saturated soils of all types can be unstable and tend to flow into the open channel. Where ditches are cut into wet land the work must be started in the driest season available and carried out in stages, by deepening a little at a time, to allow steady dewatering of the soil profile. Where channels are disrupted by spring mounds it is necessary to release the confined groundwater by separate drainage work. Where this fails the ditch must be re-routed.

Ditch excavation

Most ditches are excavated with a mechanical digger. Special digging buckets with a suitable shape for forming ditch cross-sections are available and allow direct, in-line excavation for smaller ditches but the majority are opened up with machines fitted with standard square buckets. This makes it necessary to excavate across the ditch line, and this method must be used for improvement of existing ditches.

The line of the new ditch should be marked out with pegs and excavation must always begin at the outlet end. When the natural gradient of the ditch line is a satisfactory gradient for the new channel the pegs are needed only to mark out the line of the channel since the gradient is easily achieved with a uniform depth of cut below the surface and any collecting water clears away downslope as the work progresses. Where ditch gradients have to be 'constructed' through undulations of the surface the pegs must be set out more accurately so that each peg top is above the ground level at standard height above the bed of the channel to be excavated. The gradients produced can be checked using sighting rails or other levelling instrument. The method of gradient checking by sighting rails is shown in Fig. 10.2.

Where very slight gradients are required over long distances, great accuracy is required. It is worthwhile usually to mark out the centre line with pegs and also the upper margin of both banks.

Figure 10.2   Checking the gradient

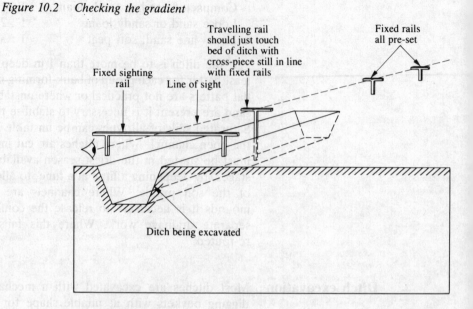

The lines of pegs marking the limit of excavation for ditch top width may not be very straight and it is easier for the operator if the pegs are joined by a rope or mark on the ground so that the start and end of each cut with the digger is easily seen. In addition to checking the gradient it is necessary to ensure that design bottom width and batter are being obtained. The most difficult part is to construct the correct batter on the bank farthest from the operator and often this bank is cut too steep. It is especially important to avoid oversteep banks where a wall or road is near the ditch as these structures can be undermined by bank slips.

Bank stabilisation   Bank stability is an important requirement in ditch construction. The majority of new channels will have stable banks if design and workmanship are satisfactory. Damage is most likely to occur immediately after excavation before a cover of binding vegetation can become established. The following few simple precautions are usually sufficient.

- The excavated material or **spoil** should be deposited at least 1 m away from the edge of the ditch. This avoids placing extra weight on the bank and reduces the chance of spoil being washed back into the channel.

- In most situations the spoil is eventually removed from the site or spread but where it is stored near the ditch, even as a temporary measure, gaps should be left at short intervals to allow free surface drainage off the land.
- Livestock should not be allowed into the ditch. Where stock may have access to the land the ditches must be protected by a sound, stock-proof fence erected at least 1 m away from the edge of the ditch. Where the ditch is the only water supply, special watering points should be constructed with a cemented drinking channel diverted from the main flow and with the channel and access securely fenced off from the main flow.

If the ditch is cut through loose sand more care is required and the problem increases with depth of ditch, increasing flow velocity and with the smallest sand particle sizes. The options listed below are in approximate order of cost and effectiveness. Choice depends on the assessed degree of bank stability and the consequences of bank collapse. If risk is high and the work expensive a qualified engineer should be consulted.

1 Spread grass seeds and fertiliser on the banks.
2 Spray the banks with clay or stabilising chemical mixed with grass seeds.
3 Line the bank with loose stone chips or gravel.
4 Line the bank with woven plastic fabric or bitumen.
5 Support the bank with piling stakes and plastic sheeting.
6 Total revetment of the banks with stones or concrete.

Culverts

Where farm traffic has to cross the ditch it is necessary to construct a **culvert** large enough to carry the greatest possible flood flow and strong enough to support the traffic. Since the lengths of piping are not great the cost of a channel large enough is not prohibitive and allowance should be made for the less frequent flood events involving, say, 30 times the average daily rainfall. The pipes used will range from a minimum of 30 cm diameter to a metre or more. They may be made of any durable material of sufficient strength – usually concrete or glazed clayware – and pipes with collars which permit positive alignment are ideal. Pipes should be laid on a firm foundation cut accurately to support the pipe walls, with cross grooves for the collars, and should form a channel that continues the line and gradient of the ditch bottom. The ditch near the outlet end of the culvert should be protected with stone or concrete to prevent channel erosion. The length of piping required depends on the width of the road, the depth of the ditch and the slope of the culvert headwalls. Headwalls can be made of cemented brickwork, blockwork or stone. To be strong enough to support the roadway they must be at least 20 cm thick with a firm foundation and the ends must be securely located in the bank by penetrating undisturbed ground by about 45 cm. Turf headwalls are suitable for smaller loads and must have a suitable slope. The pipes should be well packed and covered by at least 30 cm of stone-free material like sand or gravel so that side walls are firmly supported. The surface can then be brought up to road level by successive layers of well packed aggregate or other suitable material, each layer not exceeding 15 cm depth. It is most important to support the side walls of pipes to prevent collapse when traffic crosses the culvert. This is discussed more fully in Chapter 11.

Ditch improvement

Most areas already have sufficient length of ditches. Field drainage improvement needs a good outlet and in many cases this involves improving a nearby ditch. The improvement measures available are:

- Cleaning out and regrading the channel
- Straightening the channel.
- Resiting the channel.
- Piping and filling the channel.
- Comprehensive replanning of the ditch system.

Cleaning and regrading

Many ditches are satisfactory in all respects except that they have been neglected. All that is required is **cleaning** and **regrading** to their original dimensions. In other cases only minor adjustments are needed, like a slight deepening to allow the field drains to be at design depth or to provide a more stable batter for the banks. The design criteria are exactly those used for new channels. Faults in the original ditch design will often be evident. Where a ditch near a wall has to be deepened care is needed to avoid undermining the structure. It is necessary to move the centre line of the channel outwards to allow for the greater top width required and all excavation must be from the field side of the channel.

Ditch straightening

Where ditches or streams have unnecessary bends they may be **straightened**. As a rule channels should be as straight as conditions allow so that they are shorter and easier to maintain as well as providing the best possible channel for the flowing water. Ditches following the line of a hollow or interceptor ditches following a contour usually cannot be straightened without impairing their effectiveness. In other cases where the bends have no obvious purpose or where meanders have developed, the channel can be straightened.

The line chosen for straightening a channel should use as much of the old channel as possible. Small obstacles can be removed (by blasting out where necessary) and spoil from the new sections can be used to fill the cut-off loops. Very wide loops will require grading out since there will not be enough spoil to fill them. Care should be taken with the following points

- Where the flow is already rapid any shortening of the channel will increase the velocity and may result in erosion. Steps must then be taken to control this problem.
- Straightening a channel which follows a natural hollow can cause the channel to be moved out of the lowest ground. This is acceptable if the side slopes of the valley feature are not great in relation to the gradient of the bed of the channel. The low-lying abandoned sections can be drained by laying pipes in the hollows to discharge the water into the new channel further down-slope where it next crosses the old route.
- All associated land drains should be extended where necessary to discharge directly into the new channel. Otherwise a collector drain can be laid to connect the existing drains to the new channel.

Ditch resiting

Ditches may be **resited** to improve their drainage function or to improve land management. Ditches may have limited value as drainage channels because of faulty siting in the following situations.

- Groundwater control ditches set at the wrong spacings to achieve satisfactory control of the groundwater table.

- Collector ditches too far apart on level sites so that design limitations are imposed on the associated field drains.
- Interceptor ditches too far upslope or downslope or too far apart to control surface water flows.
- Ditches situated in areas of unstable soil where they are prone to bank slip and blockage.

Ditch spacings and layout are discussed in Chapter 15. Moving ditches to improve land management should be satisfactory if the requirements of efficient land drainage are observed.

Piping and filling

Some ditches can be eliminated by **piping and filling** to improve land management. The most common reason is to increase field size which is satisfactory up to a maximum of about 20 ha. Very large fields probably do not reduce farming costs and may introduce new environmental hazards like wind erosion. Filling ditches also releases additional land for cropping. Piped drainage channels are easier to maintain but they are less able to cope with storm flows and require a greater gradient to remain effective. This last point is very important on level sites. Piping a ditch will be worthwhile only where the drainage function of the ditch can be carried out at least as well by a buried pipe. No ditch should be piped until its function is known.

Some ditches can always be piped:

- Ditches with no land drainage function.
- Interceptor ditches to control groundflow. They are rarely successful and other drainage methods are needed (see Ch. 15).

Some ditches should not be piped apart from culverts or unstable sections or other short lengths requiring special attention:

- Ditches with gradients less than 0.1 per cent.
- Interceptor ditches to control surface flow. Pipes cannot carry out this function because surface flow passes over the filled trench. Where piping is unavoidable it is necessary to fill the trench to the ground surface with stones or broken rock to provide an easy entry for surface water.
- Major carrier ditches and storage ditches. In these cases, providing pipes of adequate size would be uneconomic.

Some ditches can be piped in certain circumstances where benefits outweigh the extra costs:

- Smaller groundwater table control ditches.
- Collector ditches.
- Smaller carrier ditches and storage ditches.

Pipe sizes, pipe-laying techniques and associated structures are discussed in Chapter 11.

Comprehensive replanning

The **comprehensive replanning** of the open channel system of an area is often required where farms are increased in size by amalgamation or where a new intensified management system is to be initiated. Such overall channel rationalisation is most needed where larger areas of flat or nearly flat land are broken up by a close network of waterways, natural or constructed, leaving the farms with small fields. In more hilly areas each natural hollow and gully must retain its watercourse whether or not the fields are too small. As already discussed, ditches can only be eliminated in the appropriate circumstances and each unit of land requires the minimum length of open channel to allow efficient land drainage at a minimum of maintenance cost. Increase of field size is possible only where there are no drainage limitations.

Planning must start with the preparation of a detailed map showing the margins of the property, field boundaries and the contour lines. Each stream and ditch should be marked in showing the direction of flow, its gradient and its function. Each watercourse which cannot be moved or eliminated must be shown in a distinctive colour since they will form the framework for the new system. Field sizes and numbers should then be arranged in a manner compatible with the management of the farm. Ditches which by chance are self-cleansing and stable and which do not interfere with the chosen field layout can be left in position since their elimination will produce no saving in costs. Other ditches which are not essential can be piped or resited as required using the methods already discussed but bearing in mind at all times the needs of associated underdrainage in the new fields.

Ditch maintenance

The majority of farm ditches contain water which flows very slowly and as such they form a good environment for plant growth. This growth provides an ideal filter for soil particles in the drainage water and the sediments built up in the bed of the channel. This and the accumulating plant remains, together with any rubbish like fertiliser sacks, raise the ditch bed until drain outlets are blocked. It cannot be over-emphasised that open drainage channels require regular maintenance if they are to remain effective, which in most cases means **every year**. Weeds may be controlled by herbicides but this is hazardous and should not be carried out without consulting the responsible River Authority. Otherwise the weeds must be controlled by cutting, usually by tractor-mounted machines. A wide range of specialised ditch-cleaning, weeding and waste-removing machines is available for efficient maintenance work. Continuing soil fertility should more than compensate for the cost involved. In all cases of ditch construction or improvement the sides of the channel should be examined for the presence of old pipe drains. Any so discovered should be improved as described in Chapter 11. In many cases this is all that is required to re-activate the existing field drainage system and no further expenditure on field drainage is necessary.

Consultations   Depending on the stability of bank-forming materials, the considerable soil movements involved in ditch construction or ditch and stream improvement may release damaging amounts of silt or fine sand into local water courses. This is particularly important where alluvium is being disturbed in a flood plain. Suspended soil particles directly affect fish by lodging in their gills, or indirectly by covering gravelly spawning grounds.

River Authorities should be consulted in advance of any such project being initiated and any conditions of consent must be carefully observed.

Chapter 11 **Underdrainage**

Most land drainage problems can be solved by installing a system of covered piped channels variously described as **land drains**, **field drains** or **underdrainage**. This last term is sometimes used to describe the whole concept of subsurface drainage including any associated soil treatments. Here it is restricted to piped, subsurface channels.

Advantages and disadvantages

Underdrainage removes excess soil water without reducing the area of cropping land or hindering field operations. At the same time, the covered channels are protected from wind-blown sediments, weed growth and the general accumulation of rubbish that blocks open channels. Because of the protecting cover of soil the drains can remain effective for many years with little need for maintenance expenditure. But when maintenance is required it is more expensive and sometimes it is impossible to reactivate an old scheme. Furthermore, the installation costs for new schemes can be much higher than those of open channel works. The other major difficulties associated with underdrainage are the absence of any overload capacity for flood conditions, the need for channel gradients to be greater than the minimum gradient acceptable for ditches and the common problem that excess soil water may not be able to flow towards pipe drains through a poorly permeable soil profile.

Types of underdrainage

Subsurface channels may be formed as unlined cavities in the soil or by setting some kind of lining material in a trench to keep a cavity open. Unlined drainage channels are formed by mole drainage techniques and are discussed in Chapter 12. Underdrainage first became widespread during the time of agricultural improvements in the eighteenth and nineteenth centuries and many methods have been tried to construct stable channels. Early types involved wedging a turf into a tapering trench or into a trench partially filled by stone rubble, brushwood or straw. Of these the stone-rubble drain was most effective since the other types became blocked when the filling material decayed and the sides crumbled. Better underdrainage was obtained by forming channels with bricks, slates, masoned stones or slotted boards. Most effective of all and most widely used were the various forms of clayware tiles manufactured for the purpose. All of these types of pipes and drain-forming materials are regularly, found, in various states of repair, when new drains are installed. The materials currently used for underdrainage are round-sectioned clayware tiles, concrete pipes and smooth or corrugated plastic piping.

Underdrainage functions

Underdrainage, like ditches, can be used for several drainage functions. The major purpose of underdrainage is to act as **land drains**, removing excess soil water from the profile to control a high groundwater table or a perched watertable. For this they must be set at regular intervals across the land. Such land drains represent the largest part of drainage work and are described as **minor drains** or **lateral drains**. Lateral drains may discharge directly into a nearby open channel or they may be connected to a single **collector drain** which may be described as a **main drain** or **leader drain**. Collector drains usually discharge into an open channel but in the larger schemes several collector drains may be connected to a large-capacity leader drain in which case the collector drains are called **submain** or **subleader drains**. Such systems of leader, subleader and lateral drains form a graded network with all drains acting as land drains to a greater or lesser extent. In some situations the leader drain may have to transport the collected water for some distance before discharging into a waterway, in which case it is described as a **carrier drain** performing the same function as a carrier ditch. Such drains do not act as land drains along the route and the pipes need not allow the entry of further water. For this purpose solid walled or sealed pipes may be used. Where leakage from a carrier drain can cause damage to nearby land it is necessary to use a sealed pipe. Where underdrainage is installed to control horizontal or upwards groundflow the channels act as **interceptor drains**. Control of surface flows is less certain and for this purpose a pipe drain must be connected to the surface with a wide band of very porous material like stones or washed pebbles. These also are interceptor drains although they are sometimes called **french drains**.

Design requirements

Underdrainage systems need to have the capacity to remove excess soil water from the profile in reasonable time. The dominant source of profile recharge is direct rainfall but there may be additional supplies from overland surface flow or from groundflow. All must be considered when selecting pipe sizes. The cost of underdrainage must relate to likely benefits and usually it is not worthwhile to cater for the worst possible flood conditons. The underdrainage must be set out in a pattern and at depths and spacings that ensure that the land is evenly and effectively drained. All underdrainage channels must have satisfactory gradients, pipe alignments and associated fixtures which together will help ensure the long-term efficiency of the system.

Drainflow

Any gravitational water finding its way into an underdrainage channel will cause **drainflow**. Where the only source of profile recharge is direct rainfall there will be a close relationship between the intensity and duration of rainfall on one hand and the volume and duration of drainflow on the other. But the relationship is influenced by other factors including the efficiency of the underdrainage system. Efficient drainage schemes with a large drainflow quickly remove excess water which ceases a few hours or a few days (depending on profile permeability) after rainfall ends while less effective systems with a smaller drainflow leave the profile saturated for a longer period continuing for up to several weeks after rainfall stops. However, if groundflow enters an underdrainage system, drainflow can persist irrespective of rainfall or soil water status.

Design drainage rate

The drainflow capacity of an underdrainage system is based on the **design drainage rate** or **drainage coefficient**. These equivalent terms are expressed as mm/day in the same way as rainfall. The design drainage rate is based on the chosen design rainfall rate but adjusted to allow for the site characteristics as affected by topography, soil type and patterns of surface flow or groundflow.

Incident rainfall can take a number of pathways, not all of which lead to drainflow. When choosing a design drainage rate these pathways must be evaluated.

Choosing a design drainage rate

Pathways

The fate of incident rainfall may be expressed as:

Rainfall = Evapotranspiration + Surface runoff + Storage + Deep seepage

These factors vary greatly from place to place and with time so that the general balance of the equation determines the natural vegetation and farming potential of an area, Storage refers to water stored in the soil and in plants but plant storage is of little significance. Where soils are saturated the plant environment can be improved by drainage and the equation becomes

Rainfall = Evapotranspiration + Runoff + Drainflow + Storage + Deep seepage

This equation cannot be transposed to provide a formula for drainflow because there are inputs of water other than rainfall. To take account of these the equation must be changed:

Drainflow = Rainfall − Evapotranspiration − Storage ± Runoff ± Groundflow.

For purposes of calculation the formula can be expressed as:

Design drainage rate = Design rainfall rate − Evapotranspiration − Storage ± Runoff ± Groundflow.

These pathways and other inflows can be considered in turn.

Evapotranspiration

On sites where incident rainfall is the only source of recharge, whenever evapotranspiration exceeds rainfall for periods longer than a few days, drainflow will cease and further loss of water will reduce the amount of water stored in the soil. In regions where the critical season is restricted to the crop-growing season, if underdrainage is needed at all, it will be used to clear excess soil water during short periods of intense rainfall which may damage crops sensitive to soil saturation. In such cases the decision to instal underdrainage, necessary perhaps for only one or two weeks each year, must be based on a finely balanced assessment of costs and benefits. In maritime regions, where there is a more obvious need for underdrainage, the critical reason usually includes the winter half of the year when evapotranspiration can be ignored as a drainflow factor.

Soil storage

Some rainfall may be needed to replenish soil storage. The amount required depends on the soil water status when rainfall starts and on the inherent storage capacity of the soil. Rain falling on land already at field capacity may become drainflow immediately. If the soil is near to its permanent wilting point, all of the rain, for a time, will be needed to top up the soil storage reservoir. The amount of water required depends on soil-storage capacity. Soils with a large proportion of mesopores – generally medium-textured soils – have the greatest storage capacity. In maritime regions it can be assumed that soils will be at field capacity for a significant part of the critical season and soil storage will have little effect on drainflow values.

Runoff

The amount of rainfall that becomes overland runoff in the form of surface flow or interflow depends on:

- *Rainfall intensity*. If the rainfall intensity exceeds the surface infiltration rate the water must remain on the surface; at least for a time.
- *Infiltration rate*. Coarse-textured soils, recently cultivated soils and surfaces with well established, undisturbed vegetation all have high infiltration rates. Fine-textured soils, bare soils compacted by field traffic, capped soils and intensively grazed leys all tend to resist the entry of rainwater.
- *Percolation rate*. Water can move freely downwards through the subsoil in coarse textured or well-structured profiles. In fine-textured or poorly-structured subsoils or where pans occur in the profile there is little or no downward percolation.
- *Soil water status*. Further rainfall cannot infiltrate into a soil profile that is already saturated to the surface.
- *Surface gradient*. If all other factors are constant the amount of rainfall that becomes run-off increases as the gradient increases.

Rainfall intensity is taken into account when the design rainfall rate is selected. This leaves only the need to assess soil properties and the influence of local topography. Gradients are very important in assessing run-off and should be measured accurately. On any type of soil, slopes of 15 per cent or more will cause near total run-off because the gradient of soil horizons converts most movements of soil water into surface flow, interflow or ground-flow. On lesser slopes the soil properties have increasing effect with less permeable surfaces likely to cause maximum run-off down to about 5 per cent gradients. Below these limits, for given values of flow rates into the soil, run-off reduces with gradient. Run-off may take the form of surface flow or interflow depending on the relative rates of infiltration and percolation, but where either factor is zero none of the incident rainfall can become drainflow, even where the surface has no gradient. To arrive at a reasonable value for runoff it is necessary to consider:

- The inherent permeability of the subsoil in terms of highly permeable, moderately permeable or slowly permeable categories;
- The surface gradient.
- Normal cropping systems as they effect topsoil permeability.
- The drainage measures to be adopted.

This last point is important where the drain trench is to be rendered highly permeable by filling it with gravel and where it is intended to disrupt the subsoil horizons. These measures have the effect of greatly increasing the percolation rate.

Run-off need not be a negative factor in the drainflow equation. Outwards surface flow on one site may be inwards surface flow on another. In many cases the surface flow is easily recognised and intercepted before it reaches the drainage site. In other cases, for example, where there is a large area to be drained with an undulating surface, interception may not be practical and allowance must be made for surface flows. For this purpose all water-spreading, water-gathering and water-neutral landforms must be identified in and around the improvement site and where such effects are likely to be significant it is necessary to adjust the design drainage rate. Major water-spreading landforms with significant run-off may not need to be drained and smaller areas may be ignored in terms of change of pipe capacity since the small over-capacity of the selected pipe size would not be prohibitively expensive. However, full allowance must be made for every water-gathering surface by selecting pipes with sufficient drain-flow capacity.

A range of adjustment factors, taking account of such vari-

Table 11.1   Likely values of r

Soil profile permeability	Drainage techniques	Disturbance of topsoil	Neutral landforms Surface gradient			Water-gathering landforms Surface gradient			Water-spreading landforms Surface gradient		
			<1%	1–3%	>3%	<1%	1–3%	>3%	<1%	1–3%	>3%
Highly permeable profiles	Pipes only	Regular ploughing	1.0	0.9	0.8	1.0	1.2	1.4	1.0	0.8	0.6
		Grass or minimum cultivations	1.0	0.9	0.8	1.0	1.2	1.4	1.0	0.8	0.6
	Pipes only	Regular ploughing	0.8	0.7	0.6	1.3	1.5	1.8	0.7	0.5	0.2
		Grass or minimum cultivations	0.7	0.6	0.5	1.4	1.6	1.8	0.6	0.4	0.2
Moderately permeable profiles	Pipes and subsoiling	Regular ploughing	0.9	0.8	0.7	1.2	1.4	1.6	0.8	0.6	0.4
		Grass or minimum cultivations	0.8	0.7	0.6	1.3	1.5	1.8	0.7	0.5	0.2
	Pipes with permeable fill and subsoiling	Regular ploughing	1.0	0.9	0.8	1.0	1.2	1.4	1.0	0.8	0.6
		Grass or minimum cultivations	0.9	0.8	0.7	1.2	1.4	1.6	0.8	0.6	0.4
Slowly permeable profiles	Pipes with permeable fill and subsoiling	Regular ploughing	0.9	0.8	0.7	1.2	1.4	1.6	0.8	0.6	0.4
		Grass or minimum cultivations	0.8	0.6	0.4	1.3	1.5	1.8	0.7	0.5	0.2
	Pipes with permeable fill and subsoiling	Regular ploughing	1.0	0.9	0.8	1.0	1.2	1.4	1.0	0.8	0.6
		Grass or minimum cultivations	0.9	0.7	0.5	1.3	1.5	1.8	0.7	0.5	0.2

(MAAF. Land Drainage Service Report No 5. 'Pipe Size Design for Field Drainage')

ables, is shown in Table 11.1. Choose the runoff factor which best fits each distinct landform, gradient and soil type and multiply the design rainfall rate by this factor. For the majority of sites a single value is satisfactory but where there are obvious variations it may be necessary to split the site into two or more subareas with separate design drainage rates.

Groundflow

On permeable, elevated sites it is likely that most gravitational water will move out of the plant rooting zone as deep seepage. Such sites are naturally excessively drained and unlikely to have excess soil water. Smaller amounts of groundwater from less permeable profiles may become deep seepage losses but where a need for drainage is identified, such losses must be insignificant and may be ignored. On the other hand, any horizontal or upwards groundflow can enter drainage pipes and alter drainflow rates appreciably. Drainflow measurements carried out at the North of Scotland College of Agriculture have shown that groundflow can cause persistent, steady flow rates that exceed rainfall by a factor of up to 6 in the observed sites. Major inflows of groundwater on this scale require separate drainage treatment but minor amounts intercepted by normal underdrainage require only a small adjustment of design drainage rate. In many cases it is sufficient to add up to 5 mm to the design drainage rate.

Siltation

It may be necessary to take account of siltation when selecting pipe sizes. Soil particles may be carried into drains with soil water and may build up to reduce the size of the channel. The particle sizes most likely to cause trouble are in the range of silt, very fine sand and fine sand. Where such particle sizes occur in a soil profile in appreciable amounts the design drainage rate should be adjusted for the more level sites to prolong the active life of underdrainage. Suitable adjustment factors (resulting in greater pipe capacity) are shown in Table 11.2.

Table 11.2   Likely values for s

Dominant soil particle size	Gradient of flatest section of pipe		
	< 1%	1%–3%	> 3%
Sand	1.0	1.0	1.0
Fine sand	0.7	0.8	0.9
Very fine sand	0.5	0.6	0.8
Silt	0.7	0.8	0.9
Clay	1.0	1.0	1.0

The most difficult profiles – composed mainly of very fine sand – may require special drainage techniques as described in Chapter 15.

Calculating design drainage rates

All of these factors discussed above provide a formula for calculating a satisfactory design drainage rate for drainage work in a maritime region:

$$q = \frac{(R \times r) + \text{groundflow}}{s}$$

where q = design drainage rate in mm/day
R = design rainfall rate in mm/day
s = the chosen safety factor for siltation
r = the adjustment value for surface runoff

Groundflow is a notional value in mm/day if direct measurement is not possible.

Pipe discharge capacity

Before this formula can be used to select pipe capacities for a drainage project it is necessary to consider the characteristics of groundflow into a pipe and drainflow along a pipe. In general terms the capacity of any covered channel to clear excess soil water depends on the rate of entry into the channel (mostly through gaps or perforations in the wall of a pipe), the smoothness of the inner surface of the pipe, the internal diameter and the channel gradient.

Flow towards pipe drains

Flow lines through an aquifer converge towards a pipe drain so there is progressive concentration of pathways reaching a maximum at the outside surface of the pipe. This restriction of groundflow is called the **concentration loss** and reduces the effectiveness of the pipe to some extent by increasing the head loss. This factor may be considered also as an **entry resistance** for groundflow into a pipe. The amount of resistance depends on the outside area of the pipe and on the size, number and arrangement of entry points in the pipe wall.

The perforations or slots ideally should be large enough to allow easy entry of soil water but small enough to hold back loose soil particles. A suitable width of slot is about 1.0 mm for most soils. In this respect good quality plastic pipes with a large number of well-distributed, clearly formed slots cause least entry resistance for a given pipe diameter while clay tiles, which allow water entry only at junctions between tiles, cause a greater entry resistance for the same diameter. Tiles with precisely machined ends can be very closely butt-jointed in position and may cause appreciable entry resistance. The rate of water entry in all types of pipes can be improved by providing an envelope of gravel round the pipe in the trench. This increases the effective diameter of the pipe in terms of water entry but a true gravel envelope is difficult to achieve in practice and usually is not worthwhile since entry resistance to the pipe is rarely a major factor in drainage failure. There are sites where slots become blocked by materials precipitated from soil water but these also are not greatly improved by gravel envelopes (see chapter 15).

Water flow inside a pipe

Once water is inside a pipe the drainflow is influenced by the effective gradient, the inside diameter of the pipe and resistance to flow along the pipe.

The effective gradient can be either the gradient of the pipe channel or the hydraulic gradient, whichever is greater. The relationship between pipe gradient and hydraulic gradient is important in terms of pipe size selection. If a field drain is set level across the field any drainflow must be due to a hydraulic gradient. If drainflow is sufficient to fill the pipe the head of water producing the hydraulic gradient must be higher than the top of the pipe channel, in which case the flowing water in the pipe is under pressure and the pipe is said to be **surcharged**. When this occurs in a level drain it is surcharged along its entire length. A pipe with a slight gradient may be surcharged everywhere except its point of origin where drainflow is not sufficient to fill the pipe. As the pipe gradient increases the length of pipe that is surcharged is progressively confined to its outlet end where drainflow reaches a maximum. Very steep drains may not become surcharged at all.

The size of pipe chosen depends on whether or not a degree of surcharging is permissible at design flow rates and it should be remembered that design flow rates will be exceeded from time to time. Barrier drains, piped ditches and the larger leader drains, together with any pipe fed directly by an open channel, must be allowed the maximum possible overload capacity and should not be surcharged anywhere at design flow rates. On the other hand, field lateral drains and most field leader drains necessarily have flows that increase steadily along the pipe. It would be wasteful to allow such pipes to be full only at the point of outlet and a compromise is possible by increasing pipe capacity (by choosing pipes with greater diameter) along its length to match increasing drainflow and allowing some surcharging at the outlet end.

Surcharging in field drains is acceptable, indeed unavoidable, on the more level sites and water will continue to enter a surcharged pipe as long as the soil water pressure is greater than that inside the pipe. As surcharging increases a point is reached when soil water cannot enter the pipe and any further increase of surcharging causes an outflow into the surrounding soil. When pressure inside the pipe is great enough the water may force a path through the soil profile and will reach the surface as a **blow out**. For this reason the amount of surcharging allowed at design flow rates should be modest. A blow out is particularly likely where a very steep drain becomes blocked or constricted where the steep gradient produces a large hydraulic head.

Pipe gradients
Ideally, field drains should have a gradient that provides, at design flow rates, a flow velocity that keeps the pipe clear of soil particles but without excessive surcharging. The great majority have flows that do not achieve a self-cleansing velocity and this factor should not influence the design unduly. As described in Chapter 15, there are sites where drains must be set level but for general drainage work, gradients of about 0.2 per cent (1 : 500) are about the limit of practical achievement, while gradients of 0.1 per cent (1 : 1000), although possible, need very precise workmanship if an even gradient is to be obtained. The maximum safe gradient for field drains is about 4.0 per cent (1 : 25) but, generally, lateral drains should not exceed 2.0 per cent (1 : 50). Extreme gradients may cause a blow out or scour the soil outside the pipe, creating cavities and pipe misalignment. Carrier drains may need steeper gradients and where pressures are likely to be great it is necessary to use solid walled pipes with sealed joints.

Resistance to flow in a pipe
The inside surface of a pipe has an inherent resistance to flow which influences the pipe drainflow capacity. The resistance to flow is called the **hydraulic friction**, which is equivalent to rugosity in ditches, and has a fairly fixed value for a given type of pipe. It is a function of the smoothness of the inner surface so that glass tubes, polished metal pipes and smooth plastic pipes are

the best water carriers. Land drains can be grouped in terms of their likely hydraulic friction value:

(a) Smooth plastic pipes show least resistance to flow.
(b) Corrugated plastic pipes show most resistance to flow.
(c) Clayware and concrete pipes show a much wider range of values depending on the quality of material and accuracy of installation but the values lie between those of smooth and corrugated plastic pipes. Well-laid pipes have equivalent values of smooth plastic pipe.

Choice of pipe size

The amount of water that flows in a pipe at design flow rates is a function of the chosen design drainage rate, the gradient of the drain and the area of land served by the drain. These values must be used to select pipe size. **The pipe sizes discussed here are the inside diameters** and care should be taken to check this value at the start because **many pipes are described by their outside diameters**. The pipe-size charts (Figs 11.1, 11.2, 11.3, 11.4) are based on the series prepared by the Land Drainage Service of the Ministry of Agriculture for England and Wales and the values chosen make allowance for all of the factors of drainflow discussed above.

Selecting pipe size for land drains – leaders and laterals

Great precision is not required and, in the case of clayware, with only a limited range of sizes available, is not possible. However, grossly oversized pipes are expensive and undersized pipes reduce the rate of clearance of excess soil water. A reasonable match is desirable in all cases. To use the pipe charts it is necessary to measure the gradient of the drain and assess the area of land served by the drain. In regular drainage systems this area is: drain length multiplied by drain interval. For collector drains the area required is the total area served by all connected laterals plus any area served directly by the collector. Irregular drainage systems need a general assessment of drain catchment area and drainflows in drains intercepting groundwater cannot be assessed by this method. For normal underdrainage, pipe discharges may be calculated according to the formula:

$$Q = 0.116 \times q \times A \text{ litres/sec}$$

where Q = pipe discharge in litres/sec
q = design drainage rate in mm/day
A = area in hectares

Figure 11.1   *Corrugated plastic pipe – field leaders and field laterals.*

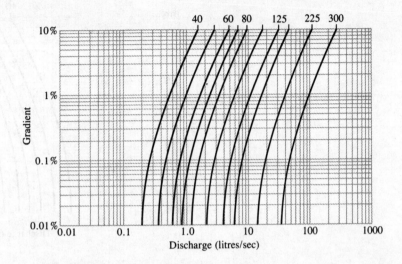

The most popular type of pipe for field drains is corrugated plastic and the characteristic flow rates are shown in Fig. 11.1. Locate the vertical line representing the calculated discharge and move along this line to where it crosses the horizontal line representing the pipe gradient. The first heavy, curved line **to the right** of this point is the pipe size needed. If this pipe size is not available the next size larger must be chosen. Where internal diameters other than those shown are available their capacities may be estimated on the chart by interpolation. In a similar way the size of smooth plastic and clayware pipes can be selected from the chart shown in Fig. 11.2.

For practical purposes, particularly for lateral drain sizes and for all except the steepest gradients, smooth plastic pipes and well-laid clayware tiles have similar flow properties.

Selecting pipe size for closed inlet carrier drains

All drains which receive inflow directly or indirectly only through the soil profile are called **closed inlet** drains and the same design criteria may be used except that no surcharging can be allowed for design flow rates. This range of drain types includes all carrier drains or piped ditches which do not act as land drains at the same time. Care is needed to ensure that all land drains discharging into the carrier are identified so that the catchment area is care-

Figure 11.2   *Clayware and smooth plastic pipe – field leaders and field laterals*

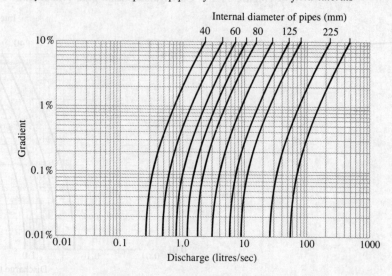

Drains which receive at least part of their inflow from open channels or from other surface flows are called **open inlet** drains and are more likely to be overloaded by flood water. The pipe

fully assessed. The pipe discharge can then be calculated. In practice it is acceptable to arrange the great range of available pipe types into two groups, those with corrugations and those without. Flow capacities for corrugated plastic carrier drains are shown in Fig. 11.3. The equivalent values for all other types are shown in Fig. 11.4. The broken line cutting across the pipe size lines on these charts indicate the conditions at which erosive flow velocities can be expected. Calculations of pipe size involving values of pipe drainflow and gradient which intersect above this line should be taken to mean that solid walled pipes with sealable joints should be selected.

Selecting pipe size for open-inlet drains

Drains which receive at least part of their inflow from open channels or from other surface flows are called **open inlet** drains and are more likely to be overloaded by flood water. The pipe capacity selected must allow for variations of surface flow as influenced by catchment characteristics and time of flood concentration. With small catchments of up to about 20 ha of only agricultural land and where no serious damage can result from an occasional flood, criteria used for closed inlet drains can apply with perhaps a longer return frequency selected as an additional safeguard if necessary. For larger catchments and where the risks

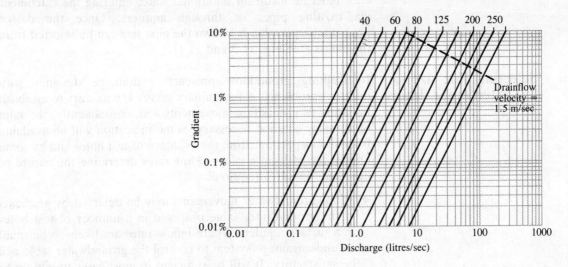

Figure 11.3   Corrugated plastic pipes – carrier drains

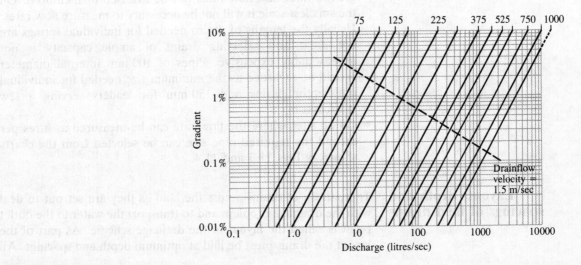

Figure 11.4   Other pipe types – carrier drains

of damage are greater a qualified engineer should be consulted. The same restrictions apply to piped ditches and in all cases allowance must be made for additional water entering the catchment via existing pipes or through aquifers. Once the design peak-discharge value is known the pipe size can be selected from the charts (see Figs 11.3 and 11.4).

Selecting pipe size for interceptor drains

Intercepting groundflow presents a drainage designer with different problems. A preliminary survey is necessary to establish the scale of groundflow movements and, consequently, the kind of drainage solution necessary. Some indication will be available from site inspection, from the evidence of test holes and by some knowledge of local geology. Flow rates determine the nature of drainage measures required.

- Minor groundwater movements may be detected by gradients in the groundwater table observed in a number of test holes in a surface aquifer. Where inflow rates are likely to be small an underdrainage system to control the groundwater table will be satisfactory. It will be sufficient in most cases to add up to about 5 mm to the design drainage rate as described on p 153.
- Major groundflows require a different drainage solution. Some evidence of the amount of groundflow or the pressure of confined groundwater will be obtained from the nature of spring issues. Test holes reaching an aquifer will release groundwater and flow rates can be assessed or measured. On the smallest scale it will not be necessary to measure flow rates. Usually the lengths of drains needed for individual springs are not great and selecting drains of ample capacity is not unreasonably expensive. Pipes of 100 mm internal diameter should be regarded as the minimum size needed for individual interceptor drains and 150 mm for leaders serving a few interceptors.
- On all sites where the flow rate can be measured as litres per second the required pipe size can be selected from the charts shown in Figs 11.3 and 11.4.

Layout, depth and spacings of field drains

The pattern of drains across the land as they are set out to deal with the drainage problem and to transport the water to the outlet point is called the **layout** of the drainage scheme. As part of the layout the drains must be laid at optimum depth and spacings. All such factors are discussed in Chapter 15 but it is essential to establish drain spacings at an early stage in the design process so that pipe sizes may be calculated. The layout must **never** be adjusted to allow for incorrect pipe sizes selected prematurely.

Workmanship and materials

Of all the factors needed to produce efficient underdrainage the most important are the qualities of workmanship and materials. Errors in the choice of design factors will reduce the effectiveness of the work to a greater or lesser extent but some improvement is obtained in most cases. Poor workmanship or inferior materials can render the work completely ineffective, either immediately or after a few years.

Drain laying

Traditionally, drains were laid by hand using a variety of types of spade and soil scrapers to open up drain trenches to the required dimensions. Good gradients could be achieved using siting rails as shown in Fig. 10.2 and gradients could always be demonstrated by pouring water into the drain trench. High labour costs have now eliminated this method of drain laying in most countries and a wide range of drainlaying machines has been developed. These machines can be grouped into three categories – excavator diggers, continuous trenchers and trenchless drain ploughs. Each type has its merits and limitations but all can do useful work if used correctly.

Excavator diggers

The mechanical digger is essentially a power-operated digging scoop or bucket mounted at the back of a wheeled tractor fitted with stabilisers. The digging arm has a 180° traverse allowing it to work behind the tractor or to either side. These machines are also known as excavator/loaders or back-acting diggers. The drain trench can be excavated to the general level required but the final adjustments of gradient, shaping the bottom channel and the pipe laying must be done by hand. The gradients must be set by the man in the trench since there is no possibility of siting a line from the tractor seat. The maximum depth of excavation is about 4 m for the larger machines and the width of digging bucket is usually about 40 cm. Bucket sizes and types are interchangeable depending on the nature of the work and the range includes narrow tapering types for economy where the cost of materials filling the drain trench is important. The most powerful excavators are track-laying machines with the digging arm mounted on a 360° traversing platform. This type of excavator is particularly useful where difficult surface conditions exist. For deep interceptor drains these machines with wider buckets can quickly and easily excavate the deepest drain trenches and provide working space for the drain layer. These machines have the height and reach to clean out ditches over the top of fences and the larger models can excavate to about 8 m. Because of their versatility, mechanical diggers are widely used and there are no limitations on the size of the pipe that can be laid. Furthermore, it is always necessary to have one on site where one of the more specialised drainage machines is to be used, as they are needed to excavate trenches at the start of each drain run, to correct faults, remove boulders or open up test holes. They are indispensible for small schemes, small undulating fields and for underdrainage to control groundflow movements.

Continuous trenchers

The continuous trencher is a self-propelled unit on tracks with cutting blades set on a chain or revolving wheel which cuts the drain trench as the vehicle moves across the field. Smaller models are available for attachment to farm tractors and are useful for minor schemes. The blades are shaped to cut the drain trench and form a groove in the bottom for the pipes which are fed into a delivery chute and automatically set into the groove. The depth of cut is controlled by the driver using preset siting rails or by laser equipment. Such large specialised drainage machines require the open fields and even surfaces of the best farming areas for economic use. Pipes of up to about 150 mm diameter can be laid to a maximum depth of about 2 m by the larger models. Choice of chain allows the width of cut to vary from 20 to 40 cm depending on the size of pipe being used with the narrow trench being economical when using trench-filling materials.

The trench remains open after the drains have been laid, allowing inspection of the pipe and the identification of lines of old drains which may then be connected to the new drain as required. Where necessary the spoil can be allowed to dry out on the surface before backfilling. The machine can be used for a wide range of soil conditions but profiles containing many small stones cause rapid wear of the blades and large boulders cause the whole cutter unit to ride out of working depth. Loose soil may sometimes fall back into the trench before the pipes are laid, causing a soft, uneven bed for the pipes. Care is needed to avoid this. A continuous trencher is probably the most accurate machine for work on the flatter sites.

Trenchless drain ploughs

Trenchless drainers essentially are very large mole ploughs with a delivery chute which plough pipes into position in the soil profile. Most are very large, self-propelled, track-laying vehicles. They are best suited to larger drainage projects and the more permeable soils which can provide suitable working conditions for a greater part of the year. They are restricted in heavy soils because dry conditions are needed if they are to operate satisfactorily. Pipes up to about 150 mm diameter can be laid to a maximum depth of about 2 m using the very largest models. Depth control is as described for trenchers and can be accurate in good conditions but attempts to drain in wet conditions, in variable or very hard soils or at too great a forward speed, can lead to poor gradients. The narrow slit created by the passage of the machine makes it the most economic means of installing permeable materials above the pipe and there is much less wear on the underground parts so that machine maintenance costs are much less than for trenchers. Large stones cause difficulty since any attempt to reverse the machine or to lift the blade will distort the pipe already laid. In the best conditions the passage of the blade effectively disrupts any induration in the profile and thus improves the permeability. In wet clay profiles the blade smears the sides of the fissure as it passes, creating a vertical barrier which impedes flow towards the drain. Perhaps the most serious deficiency is the absence of an open drain trench to allow examination of the new pipe and to give access to connect old drains where necessary. In best conditions, however, drains and permeable materials can be laid quickly and accurately at less cost per unit length than any other machine.

Choice of pipe

The choice of pipe for land drains is largely a matter of availability, cost and personal preference. There are some technical constraints which need to be considered. Tiles are made 30 cm to 45 cm long, smooth plastic pipes in 6 m lengths, while corrugated plastic pipes are supplied in rolls containing from a few to 150 m of pipe depending on the pipe diameter. A spool of continuous pipe is particularly suited to specialised drain-laying machines because the pipe feeds from the spool through guides and chutes directly into the required position in the soil profile. Other types of pipe must be fed into the chute by hand. In all types of plastic pipe the ends of pipe lengths are fitted together by collars so that misalignment is most unlikely. The greater range of available pipe diameters permits a very close match of pipe size to design flow rates. The very light plastic material allows easy transport of pipes across wet land or difficult terrain during the drainage operation and soil structure damage is less likely. On the other hand, there can be disadvantages associated with the use of plastic pipes. Pipes manufactured from polyvinyl chloride (PVC) become very brittle at low temperatures and fracture easily so that they cannot be handled unless the ambient temperature exceeds 5°C. Pipes made from polyethylene or polypropylene can become soft in hot sunshine and may distort when handled. Furthermore, plastic pipes have been known to collapse under pressure of overburden when laid more than about 1.5 m deep. Most operators prefer the sturdier, more manageable clayware tiles for deep interceptor drains. Large coils of pipe are awkward in a deep trench and collecting groundwater may not be able to enter the pipe fast enough through entry slots to prevent flooding of the working area. The always open end of tile drains avoids this difficulty.

Choice of pipe for leader drains and carrier drains

A greater range of pipe types is available for large capacity drains. In addition to those described above there is available a variety of non-perforated plastic pipes, glazed clayware pipes, concrete, asbestos cement and other types some of which are of reject quality from the very demanding specifications of pressure pipelines and sewerage work. All can be satisfactory if they are of appropriate quality and dimensions for the particular function required on the site. Where the pipes must act as land drains the walls must be perforated or short, butt-jointed types must be used. Porous concrete pipes, designed to allow water to pass through the pipe wall, are also available for the purpose. Where a drain is to act only as a carrier, any pipe of sufficient strength and capacity is suitable. Sealed pipes are needed where high flow velocity is expected and spigot and faucet sewer pipes are ideal for the purpose. Where very large pipes are needed the appropriate size of concrete pipes are most often selected. These have the necessary capacity and load-bearing strength. There are some limitations in the choice of concrete pipes. Where subsoils are more acid than about pH 6.5 or where the groundwater contains more than about 0.06 per cent dissolved sulphate, ordinary Portland cement is corroded and the pipe life is shortened. Porous concrete is particularly susceptible and the pipes can disintegrate after only a few years. In case of doubt the soil should be tested and more resistant pipes used if necessary. Sulphate-resistant cements may be specified where concrete is essential for strength purposes.

The quality of materials

The materials used for drainage work should be of a type and quality that performs the required function in the particular location for a sufficient number of years, often 50 or more, to make the investment worthwhile. Clayware tiles should be made from good quality, well-fired clay and have strong uncracked walls free from distortions. The ends need to be neatly trimmed to permit good close butt joints. To avoid frost damage the tiles have to be kept dry when in storage. Plastic pipes must be made from good quality material and be of adequate strength for land drainage work with well-formed slots in sufficient numbers along the pipe. PVC pipes must not be handled in temperatures less than 6°C. The plastic material can also be rendered brittle by prolonged exposure to sunlight so that old, bleached stocks should be discarded. Most countries have agreed minimum-quality standards for pipes for land drains and only pipes meeting these standards should be used. All builders' materials needed for ancillary fixtures and fittings should comply with minimum building standards and regulations.

Suitable conditions for pipe laying

The best quality land-drainage work is achieved when the soil profile is dry, a condition that occurs mainly in the summer half of the year in maritime regions. This is worthwhile even at the expense of losing a crop although the loss is not necessarily inevitable because experience of the technique of 'through crop' drainage has demonstrated that yield losses in the year of installation can be surprisingly small. Drainage at other times of year is quite feasible in naturally permeable soils, during dry spells and, in less ideal conditions, where the drain trench can be left open until the profile and spoil has dried out. In profiles that contain a high proportion of clays and silts it is essential to avoid working in wet conditions which may incur several problems – the work is slower and more expensive; machines may be delayed or bogged down; accuracy is lost as machines wallow or crab on slippery surfaces; and soil structure is damaged by the pressure of heavy drainage machines rendering the profile less permeable. Perhaps most important of all is the loss of permeability near the drain trench caused by the smearing action of the cutting surfaces of the machines. This can be cured, to some extent, if the trench can be left open until the exposed soil dries but trenchless techniques pose a difficult problem in wet clay profiles. In the worst conditions, with any type of machine, slurried clay placed above the pipe can invade slots and seal off the pipe. Some sites – spring issues and wet hollows – are never dry and these must be drained in wet conditions. Only excavator diggers can be used for this work. Drains may be laid immediately or, in the very worst conditions, the site may be dewatered by open channels as a first step.

Setting out the scheme

Before any drain trenches are opened the siting of all drain lines should be clearly marked out with pegs. In turn the outlet point, the leader, subleaders and laterals should be carefully set out. For straight drain runs a peg at each end is sufficient. For leaders following a hollow or interceptors influenced by land contours the pegs need to be close enough to clearly indicate the required drain line. For work to be done by excavator diggers on slight gradients or undulating surfaces it is advisable to indicate the depth of cut needed at each peg. A marking peg is needed at each change of surface gradient and all peg tops must be set at a constant height above the required trench bottom. This kind of site preparation can only be done using an accurate surveying instrument.

Drain gradients

It is essential that all drains be laid with an even gradient to maintain a steady or increasing drainflow velocity towards the outlet. If flow velocity is reduced along the pipe by reduction of gradient, inadvertant sags in the drain line or by swirls of water at drain junctions, siltation can occur and the pipe may be blocked.

Achieving accurate gradients

The amount of care needed depends on the surface gradient. Where excavator diggers are used to open the trench, the pipes are laid by hand and the gradient can be checked as work progresses. Where the drain line has a pronounced gradient the flow velocities will be relatively fast and it is sufficient to check the general gradient in relation to an even surface or simply by laying a straight edge and spirit level along the tops of the line of pipes as work progresses. On lesser slopes the sighting rail method illustrated in Fig. 10.2 can be used and for the most accurate work the gradients must be checked by surveying instruments. With drain laying machines the depth of cut is controlled by the driver as the machine moves along the drain line. He constantly adjusts the height of his control platform (to which the digging mechanism is rigidly attached), keeping his sighting device in line with pre-set rails at the end of the drain run or in response to signals from laser-levelling equipment. Whichever method is used the initial setting up is crucial to the success of the whole operation. Setting up laser-levelling equipment is relatively simple except that the transmitter must be firmly fixed so that it does not waver in the wind and the chosen gradient is accurately set following the makers instructions. The transmitter unit is set on a tripod and revolves to form a plane of light which can be tilted with great accuracy to the required gradient and one setting can be used for a number of drain runs. A receiver unit mounted on a mast on the drainer senses the laser beam and indicates to the driver by means of coloured lights on the control panel whether the height of the receiver (and the depth of cut) is correct, too high or too low. Most equipment also permits automatic depth control by means of electrical signals from the receiver which directly activate the depth control mechanism. Where the topography requires different gradients along parts of a drain line this can also be controlled automatically. All that is necessary is an additional gauging wheel to give the control unit a sense of distance travelled and, when so programmed, the receiver unit is progressively and precisely raised or lowered on the mast to create a drain gradient that differs from the laser gradient by the required amount.

The difficulties of gradient control

With skilled operators, well-maintained equipment, accurate setting up and good working conditions modern machines are so accurate that they have widened the scope for piped underdrainage. But the work is not always done well. The initial gradient setting of the laser beam (or sighting rails) is the key to success and must be done by an operator with sufficient training and experience. It is not unknown for the laser to be set at the wrong gradient and for the driver to make a series of abrupt depth alterations as the machine comes to the surface or digs impossibly deep along the run and the drain is laid as a series of steps! Even with accurate laser setting troubles can occur. All depth-control mechanisms for cultivating machinery depend on the bite and penetration of the digging parts countered by the lift of hydraulic rams. A control unit seeks to balance these opposing forces at a chosen level in the soil and can do so if the rate of forward travel is compatible with the speed of reaction of the control mechanism and if the soil is uniform. As speed increases and/or the soil becomes more variable the lag in depth correction increases. Even with correct speed settings the more difficult soils may cause inaccuracy. The heavy digging mechanism may sink into soft patches or it may ride out of working depth at rocky outcrops, over boulders or where extreme induration occurs in the subsoil, despite attempted corrections by the depth control machinery. With trenchless machines in particular these are real hazards. A further difficulty is that the blade always tends to bite too deeply at the start of a drainage run before the depth controller takes effect, resulting in a sag at the outlet end of the drain. Experienced operators are aware of all such shortcomings and make allowances as necessary and identify all points of unavoidable loss of depth control to allow later correction using a digger. In the most difficult sites it is better to use a digger for the whole scheme. In all cases it is wise to check pipe gradients independently to ensure that the work is accurate.

Drain outlets

All drain laying runs must start at the outlet end and progress upslope. This ensures that each drain is at correct depth at its outlet and removes collecting groundwater from the working area. Lateral drains must start at the same level as the collector drain and all pipes with outlet to an open channel must be set at lease 15 cm above water level at design flow conditions. Where this is not possible the outlet must be at least 15 cm above the channel bottom to minimise the chance of pipe outlet burial.

The exposed outlet end of a drain is vulnerable to damage and disturbance. The first 1.5 m or so should be solid walled and frost proof (that is, not of unglazed clayware) and the first 2.0 m of the trench must be tightly packed with spoil to secure the pipe and to reduce the chance of leakage from the pipe which may result in soil erosion and pipe dislodgement. Where each lateral drain has direct outlet the pipes should project beyond the bank to reduce bank erosion but not more than about 50 cm since the weight of ice and snow may cause plastic piping to collapse. Telescopic or removeable end-pieces are useful to allow bank maintenance and, where banks might be easily eroded, plastic chutes are available for fitting below each drain outlet.

The outlets of leader drains need to be substantial and conspicuous constructions with a frost proof outlet pipe set into a headwall of cemented stones, blocks or bricks or into a substantial prefabricated slab strong enough for the purpose. All types should have a dripstone to prevent channel erosion. Pipes with diameters greater than 6 cm are better fitted with a metal grid to keep out animals which may enter the pipe and die, causing drain blockage. Pipes may be blocked also by plant roots. To minimise such damage all drains within 5.0 m of a hedge or tree should be of solid-walled pipes with sealed joints.

Drain depth

Field drains must be at design depth for most of their length. Little harm is done where an even gradient carries a drain to greater depth through a minor surface undulation but where drains are to run through a larger elevated section, the gradient or layout must be adjusted to achieve the optimum drain depth. Very shallow drains are easily dislodged by cultivations or crushed by field traffic so a minimum cover of 60 cm is essential for pipe protection. Some discretion is possible through limited low areas provided that extra soil can be carted in to provide the necessary cover. In a similar way the first few metres of a leader at a difficult outlet point can be left too shallow provided that the section is permanently fenced off.

Pipe foundations

Drain pipes must be set on a firm foundation. Soft or unconsolidated soil should be avoided because settling or shrinkage after drainage will cause pipe dislodgement. An unconsolidated foundation is created where a digger is allowed to cut the track too deeply and the correct gradient is made up by shovelling back loose spoil. Also, loose spoil may fall back into the trench behind the cutting chain of a trencher creating a loose uneven foundation for the pipes. Care is needed to avoid both problems. Where only a short length of unstable ground must be crossed, perhaps an old ditch line or a recently backfilled excavation, the pipes can be kept in alignment by laying them on a securely founded bridging device like a reinforced concrete beam or iron pipe resting on undisturbed soil on each side of the unstable ground. Where necessary, concrete footings can be set in to support the bridge.

Deep peats are troublesome because of uneven shrinkage and the traditional practice of laying tiles on boards is not really satisfactory. Plastic pipes retain alignment better and gravel fill above the drains permits continuation of flow even where the pipe sags.

Support for pipe side walls

When a drain trench is backfilled the overburden exerts a pressure on top of the pipe. To prevent pipe collapse it is necessary to support the side walls by setting the pipe in a groove which snugly fits the outside diameter. The groove is formed automatically by trencher and trenchless machines but where the trench is opened up by a digger the groove must be cut by hand tools. Greater care is needed for deep drains than for shallower drains and for plastic pipes rather than clayware pipe.

Where pipes are used in culverts or at any other point where heavy loads may cross it is necessary to select pipes of sufficient strength and the pipe supplier or an engineer should be consulted about the type and quality of pipe required. The pipes should be set in a clearly formed trench and then supported by well-packed, stone-free filling material. Transverse grooves must be cut to take overlapping pipe collars and allow the whole pipe to rest on the floor of the trench. Where the channel bed is rocky or stony the pipe should be set in a bed of stone free material like sand so that projections of rock do not press against the pipe.

Pipe laying

All types of pipe requiring butt joints must be located securely in the pre-formed groove and butted closely together end to end to provide a smooth continuous channel. Where tiles are laid by a draining machine they are fed into a chute and deposited in place as the machine moves along. Care is needed to ensure that the pipes slide smoothly through the chute and lie closely together in position otherwise they will become blocked very quickly. Hand-laid tiles are usually well set together, and when using normal tiles with fired ends there are always small irregularities which prevent a very close fit; but when tiles with machined ends are used they can fit so closely together that the drain has a high entry resistance. With this latter type it is better to use a 1 mm gauge to create a suitable slot. Plastic pipes are connected by collars so that good joints and pipe alignments present no problems, but care is needed at all junctions and joints between pipe sections. Plastic piping is very light and is easily dragged out of the socket when the machine moves forward. A second operator is needed to secure the pipe in position.

Connecting new drains

All junction points between drains permit a smoother flow of water if they are constructed by means of purpose-made tile or pipe-junction pieces. A range of types is available allowing connection of most sizes of lateral drains to most sizes of collector drains and at different angles of alignment to a maximum of a right-angled junction. The branch connections are set high up on the shoulder of the larger pipe so that the main flow is unimpeded; for the same reason junction angles of more than 90° must be avoided. Any serious loss of flow velocity at pipe junctions can cause siltation. Major junctions between carrier or leader drains should be constructed as inspection chambers complete with a silt trap. An inspection chamber should consist of a suitable concrete floor with walls of brick, pre-cast concrete or vertically set concrete pipes, with a lid. The chamber itself should be of sufficient size to allow easy cleaning of the silt box and should be strong enough to be safe and stable. Advice should be sought on the standard of construction required.

It is usually better to construct the inspection chamber with its lid above ground level where it is easily located for maintenance work. Buried inspection chamber covers cause less disruption of field operations but are easily lost or forgotten. A similar type of construction is very useful at all unavoidable points of decrease in gradient or at sharp corners to allow regular removal of the accumulating silt. The silt box should be at least 30 cm below the level of drainage channels and the silt should be removed before it builds up to the level of the channels, otherwise its effect is lost. To prevent water backing up in the drains, the outlet pipe should be the lowest pipe in the inspection chamber.

Connecting old drains to the new system

Most land in the old farming areas of the maritime regions has been drained at least once in the past. Many of these old under-drainage systems are at least partly functional still or are rendered functional to some extent by the new drainage work. Most have long established connection to the topsoil and rapidly fill with gravitational water. On sites with appreciable slopes it is often the blockage of old drains that causes wet patches and the drainage may not be improved by new work unless the old systems are effectively connected. There are some cases where it is *not* necessary to connect the old drains:

- Sites that have very permeable soils.
- Level sites where the old drains are deeper than the new ones.
- Sites where the old drains are very shallow and can be disrupted easily by subsoiling.

In all other sites where old drains might be functional they should be connected to the new system and for this purpose the presence, depth and condition of all old drains should be established before the new work begins. As work proceeds all open-drain trenches should be examined for old drains and each one on the upslope side should be connected to the new drains by junction pieces or by placing enough gravel in the trench to provide a good connection between the drains. Where the old drains are deeper than the new system, occassional new drains, at strategic points such as the bottom of a slope, should be laid sufficiently deep to cut and drain the old system. Where trenchless machines are used the only possible solution is to provide a continuous band of gravel in all trenches to connect old drains. Even this may fail in fine-textured soil because the passing share tends to push tiles aside and plug them with clay. Careful observation of the site throughout the next critical season is necessary to identify old drains still causing wet patches. These points must be excavated and effective pipe junctions constructed.

Backfilling a drain trench

Pipes must not be left exposed in a drain trench after installation. If gravel trench fill is not to be used the pipes should be covered immediately by a layer of stone-free soil, not less than 15 cm deep, placed carefully but firmly in position to give protection from frost damage and from damage or dislodgement when the trench is filled. The spoil should be left on the surface until it is dry and all large stones should be removed. The final trench filling is usually by angled blade on a tractor and no attempt should be made to compact the trench fill.

Piping ditches

The general principles of drain laying apply to piping ditches but some points need special attention. The line chosen for the drain may be the bed of the ditch but not necessarily so if the route is not ideal or if the bed is soft. Whatever route is chosen it is essential that all field drains discharging into the ditch are located and connected to the new pipe, either directly or via intermediate collector drains. Where unavoidable gradients will cause drain-flow velocities in excess of 1.5 m/sec, as shown in the pipe-size charts, all pipes must be sealed. There should be inspection chambers at all major pipe junctions, sharp bends and changes of gradient. In certain cases only part of a ditch is piped and the pipe must accept water from an open channel upstream, which involves accurate pipe size selection. It is necessary to guard the open inlet drain from blockage by sediments and floating debris. A suitable type of silt trap and protective grid resembles an inspection chamber as described above but with one wall replaced by a stout metal grid set at the end of the open channel. It should conform to the standards described for an inspection chamber. Discharges of waste liquids from farm buildings should not be allowed to enter land drainage pipes since the waste may block the drains or it may pollute a local water course.

Pipe maintenance

Pipe drainage systems need very little maintenance but they cannot be forgotten completely. Outlets in open channels need to be kept clear and silt traps must be cleaned out regularly. The pipes may need some clearance from time to time as sediments or other materials collect, and it is useful to have an accurate plan of the scheme. Scale plans should be prepared for all new drainage schemes and carefully stored for future use. The traditional method of cleaning field drains is to open up pits at intervals along each drain, remove a tile and push a sectional flexible rod fitted either with brushes or scrapers along the drain. Drain jetters now do this work more easily and can reach from the outlet or junction of each drain to the far end. A pressure hose fitted with a special jetting nozzle propels itself along the pipe and the swirling water flushes out sediments as the jetter progresses. Both systems are much easier where the drains have individual outlets but all layouts can be improved as long as the pipes can be located and opened up.

Drain blocking will be noticed as progressive deterioration of soil drainage status or as patches of wet land appearing where none occurred before. Investigation of the problem must be carried out in a systematic manner. The drain outlets should first be examined to make sure they are clear and remedy any obvious faults. If this is not sufficient it is necessary to open up some of the drains, usually choosing a wet area if the wetness is patchy. Exposing the pipe will reveal one of two conditions: either it will be full of water and water from the pipe may even flow into the pit, or the pipe may be empty with wet soil above. In the former case the pipe system has become blocked downslope from the test hole and it will be necessary to clear the blockage. Silt may be jetted out, but if blockage is caused by a collapsed pipe or poor connection the problem section must be located. The simplest method is to push draining rods along each section in turn until the obstruction is found.

When the pipes are found to be empty there are three possible reasons: the pipe entry slots may have become blocked by slurried clay, ochre or peat fibres; the drain trench-filling material may have become impermeable; or there may be general impedance in the soil profile; for example, by a cultivation pan. Of these only the last is easily remedied by subsoiling. The others usually require installation of a new drainage scheme.

Chapter 12 **Subsoiling, mole drainage and permeable fill**

Many soil profiles are either totally impermeable or are so slowly permeable that there is no significant downwards groundflow. In such cases underdrainage alone usually fails to solve the problem of excess soil water. When this occurs it is necessary to seek to improve subsoil permeability by some type of disruption and to ensure that the material filling the drain trench remains permeable.

Perched watertables In slowly permeable soil profiles, excess soil water remains near the surface as a perched watertable. Underdrainage in such a profile will clear the excess water as long as the disturbed soil in the drain trench remains permeable but there will be very little groundflow through the undisturbed subsoil. Should the soil in the trench become impermeable, the drains, however closely the lines of drains are set together, lose contact with the surface and cease to function. When this occurs excess soil-water remains at the surface or flows over the surface into nearby hollows.

Soil treatment A perched watertable always indicates impermeability in the soil profile if it is more than transient and associated with recent heavy rainfall. The impermeability may be caused by a pan, by general induration, by a massive clay subsoil or by a combination of these factors. Examination of the profile will identify the cause and indicate a remedy. The drainage solution must include some form of **soil treatment** to create macropores in the general profile. The options available are to form unlined underground channels called **mole drains**, various forms of soil cultivation including

subsoiling and the technique of placing permeable material – normally described as **permeable fill** – in the drain trench.

Mole drainage

Where the profile impermeability is caused by a dense clay subsoil the best means of improvement is mole drainage. Mole drains are unlined channels created in the subsoil by pulling a mole plough through it which creates a cavity in the soil without the need to open up a trench. The method depends on the plastic, cohesive nature of clay and is effective when the work is well done. It is also much cheaper than closely spaced underdrainage. The disadvantage is that the unsupported channels will collapse sooner or later and in most cases must be renewed in a routine manner.

Mole ploughs

A mole plough creates an unlined channel in the subsoil as it is pulled across the field. The mole consists of a solid, circular section, steel rod about 7–9 cm in diameter, pointed at the front to give penetration and trailing an expander piece behind, which is also cylindrical and slightly larger than the mole, to give the channel its final shape. A vertically set blade or tine attaches the mole unit to the frame of the implement and allows the selection of a range of depths of working. On the larger trailed machines the blade is attached to an iron beam up to 3 m long which slides along the ground surface and smoothes out the mole channel gradient over minor undulations. The sliding beam causes a good deal of friction as it moves along, greatly increasing the power and traction needed. Smaller machines are available without beams and are attached to the linkage of normal farm tractors. This type of machine is more sensitive to surface variations and to tractor pitching movements but is much easier to pull.

Conditions needed for good mole drainage

The most favourable conditions for mole drainage are homogeneous, clay-rich soils at the optimum soil water status. The very outside limits of texture that can be considered for moling is a clay content of not less than 25 per cent although most authorities would say that 30 per cent of clay is the minimum. A good, but perhaps over severe, test is to shape a piece of moist subsoil into a ball and place it in a container of water. If after a day or so of immersion the ball has not collapsed the soil is very suitable for moling. The soil water status is also important since dry clays are hard and brittle while moist clays are soft and plastic. Good mole drainage requires a stable, well-formed channel connected to the surface by fissures. The best conditions occur when the soil around moling depth is in a plastic state and the soil above is in a friable state. When moled in this condition a clean smooth channel and slot are formed and the passage of the blade heaves and fractures a cone-shaped zone above to increase the soil permeability to a marked extent. A cross section of this effect is shown in Fig. 12.1.

Figure 12.1   Ideal mole channel

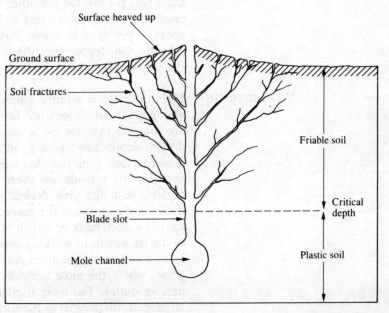

When the soil is plastic to the surface only the slot is opened up above the mole channel and this is quickly closed by slurried soil and field traffic. When the soil is dry to mole depth the mole itself causes fracturing so that the roof of the channel collapses. A drying-out phase of weather is better than a wetting-up phase. Should continued soil water loss dry out the zone around the newly formed mole channels little harm is done, but if the channel walls are saturated soon after formation the recently disturbed clay particles have lost some of their cohesion and rewetting may induce exceptional instability that will cause channel collapse.

Mole channel gradients

The depth of work of a mole plough is set in advance and cannot be altered when the machine is in motion if it is a trailed type. In mounted types the depth of work can be altered but there is no means of gradient control. The channel gradient is determined by the slope of the surface. For general work the minimum acceptable gradient is about 0.5 per cent since, on lesser slopes, lack of gradient control will leave water standing in depressions in the channel with the saturated soil causing channel collapse. On very flat sites moling is possible only if associated with relatively close set collector drains so that each length of mole channel is short (see p 179). On the other hand, unlined channels are easily eroded, making it necessary to limit gradients to a maximum of about 6 per cent in stable clays or about 4 per cent in other subsoils. On steeper sites the mole runs should be angled across the slope to be within the best range of channel gradients.

Workmanship

Mole drainage is a fairly simple operation with very little to go wrong when conditions are favourable. The basic requirements are that the machine be in good condition and well set up, the design depth and spacing adhered to and that good steady gradients are achieved. Sufficient power and traction must be available to provide an even pull upslope from the channel outlets. Just like pipe drains, the mole run must begin at the outlet end to provide the correct depth and gradient. Runs may start in a ditch bank or trench or the mole plough may be allowed to dig its way in to working depth from the surface. In the latter case, a collector pipe drain must be located along the line of the points where the mole achieves working depth to give the channels an outlet. The mole itself must be in good condition, truly aligned to run parallel to the surface and in line with the direction of travel. Insufficient 'bite' will cause the mole to ride out at hard sections, while a mole pointed downwards will travel at an angle to the surface and will tear a ragged oval-sectioned channel which lacks stability. In the same way a bent blade causes the mole to crab also producing an oval, ragged channel. Care is needed to ensure that a clean circular sectioned channel is being formed at the correct depth and this can be inspected only by exposing the soil profile.

Design factors for mole drainage
Ideal clays for moling

Some clay soils are well known for their good moling qualities. Usually these are homogeneous, calcareous, stone-free clays of the smectite group of minerals that are subject to an appreciable soil water deficit in the dry season. Soils of this type can be drained effectively using mole channels only.

Where gradient permits, mole runs up to 200 m long are quite feasible but for such long runs it is essential to use only heavy machinery with beam mole ploughs to obtain a good even gradient. A suitable working depth is about 60 cm and channel spacings can be from 2.5 m to 9 m depending on rainfall and known drainage characteristics of the site. The mole channels may need to be renewed every 7 years or so but usually they last longer, sometimes much longer. However, where this happens the life of the channel may be limited by outlets, which easily become blocked. It may be worthwhile laying a short length of piping at each mole outlet or, better still, to lay, in advance, a pipe collector drain to act as a common outlet for the mole channels.

Moderate clays for moling

Most clay soils are less than ideal for moling. This category may include soils with the minimum clay content or the more acid clays or the less stable minerals like the kaolinite group, especially in areas where the soil water deficit is smaller. Other limitations are uneven surfaces, pockets of sand within the clay profile and stones in appreciable numbers. Most of these soils can be mole drained quite successfully if a system of pipe-collector drains is installed before the moling begins. The collector drains are the key to success and must be set out across the slope as much as the gradient will allow at a depth that provides an easy outlet for the mole channels. Generally the top of the collector pipe should be at least 15 cm below the bottom of the required mole channels. The drain trench must be filled with enough permeable fill to carry the permeable connecting zone to at least 15 cm above the mole runs. The spacing between the collectors is a function of the likely stability of the mole channels and normally should be in the range of 20–40 m. Where the slope is uneven, moling is still possible by siting collector drains to run through hollows and along lines of change of slope. To avoid water backing up along the channels in wet weather the collector drain discharge capacity should be based on one-day storm rainfall values. Collector drains must be laid first; their presence helps

Figure 12.2   Moling through the permeable fill

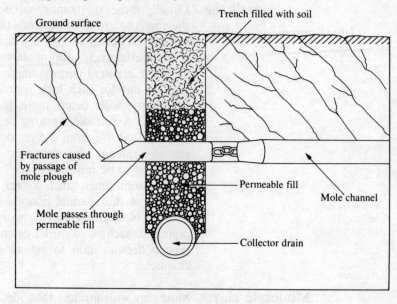

to clear surface water, allowing suitable moling conditions to develop more rapidly as the soil dries out. When the soil profile has dried sufficiently the moles should be drawn upslope, across the lines of collector drains and through the permeable fill as shown in Fig. 12.2.

The direction chosen for moling runs should provide the shortest possible length of mole channel between collector drains that that is consistent with the gradient limitations of the mole channel. The moling depth is usually from 45 to 60 cm, with the lesser values used on the wetter sites. Mole run spacings range from 1.0 to 2.5 m, depending on the assessed profile permeability and stability. The area of land drained by each mole channel can be as little as 1.0 m × 20 m so that if a few moles collapse because of pockets of sand, stones or patches of liquid clay their loss has a minimal effect on the whole scheme. It can be expected that remoling will be required after 4 to 7 years, certainly after the first moling. The smaller, tractor-mounted mole ploughs are quite suitable for this type of work.

Poor clay soils for moling

Clay soils with the minimum clay content containing a high proportion of silt or fine sand, acidic kaolinite clays in wet areas and illitic or other micaceous clays are generally poor soils for moling. They tend to pass into a slurried condition when saturated so that any cavities in the soil may be lost after only one wet season. In fact the least stable clay soils in the wetter areas cannot be comprehensively drained. All that can be achieved is the removal of surface water as described in Chapter 13. Many, however, can be more comprehensively improved and the options available are discussed below in the section on subsoiling.

Moling other soil types

Soil with less than the minimum clay content cannot be mole drained because there is insufficient cohesion between soil particles to support a mole channel. However, organic soils do possess a degree of cohesion and moling has been found to be a useful method of dewatering peat as a preliminary to underdrainage. The initial peat shrinkage causes very little harm because the channels are easily redrawn.

The long-term effects of moling clay soils

In dense, fine-textured subsoils the water is firmly held and there is very little gain or loss in the profile below the zone of root penetration which can be quite shallow. The effect of the first-ever moling operation is to form cavities in the dense subsoil which allows a degree of dewatering of the soil mass around. The dewatering is by diffusion only and is slow but rewetting is equally slow and further rainfall, instead of collecting in or on the soil as before, can enter the drainage system quickly and effectively bypass the soil profile. Thus, year by year, as long as the mole channels operate, the dewatering process in the soil mass becomes more effective and natural structures develop. As moling is repeated over the years, especially if associated with good husbandry techniques, the natural permeability so increases that no further moling is required. The collector drains become quite adequate as land drains with perhaps additional drains installed to provide a 20 m spacing. Even in the poorer clays some dewatering is achieved with the first moling. It is commonly found that the channels collapse after one season, but provided that the moles are redrawn in the next dry season, there will be a gain in soil stability and the second set of channels will last longer.

Subsoiling

Subsoiling is a general term describing any type of soil cultivation that is intended to break up the soil horizons beneath the cultivated topsoil. It is not always a drainage operation – it may, for example, be needed to allow plants like carrots or sugar beet to form good tap roots – but only the drainage aspects are considered here. For drainage work it is often a once-only operation or it may need to be repeated at long intervals or once per rotation of crops depending on the need to improve soil permeability. The basic function of subsoilers is to disrupt any layers of compaction or induration that may occur in the profile, rendering it less permeable than the inherent permeability of soil of that particular texture. In general subsoiling is more suited to the non-clay soil textures. Shallow impermeable layers may be broken up by a range of cultivating machinery but the deeper ones require a subsoiling implement designed for the purpose.

Subsoilers

The basic features of a subsoiler are similar to those of a mole plough except that the digging point or shoe has a different shape and a sliding beam is not required. The digging shoe is flattened and chisel-like, designed to lift and shatter the soil as it passes. If the tine is fitted with wings set at an angle to the horizontal a wider band of soil is lifted. Smaller machines have a single tine which is fitted directly to the tractor linkage while the heavier types can have two or more tines fixed to a heavy frame for attachment to a powerful crawler tractor.

The effects of subsoiling

Subsoiling can improve profile permeability only where the soil is in a hard or friable condition. In this condition the passing tine makes way for itself by lifting a zone of soil, so creating a network of fissures all the way to the surface. A cross-section through a zone of shattering is shown in Fig. 12.3. If wings are fitted to the blade a much wider zone of shattering occurs, but the penetrating effect of the tine is reduced. In very hard profiles there is resistance to the penetration of any type of tine and attempts to subsoil with machines of inadequate weight and power merely form grooves on the upper limit of the zone of induration within the profile. Attempts to subsoil when the soil is in a wet plastic condition have a quite different effect. The tine makes way for itself by deforming the soil (by squeezing it) so that no profile shatter is achieved. A rough, unstable, square-section mole channel is formed which soon collapses. The likely results of poor quality subsoiling are illustrated in Fig. 12.4.

Coarse-textured soils are nearly always friable, medium-

Figure 12.3   Ideal subsoiling

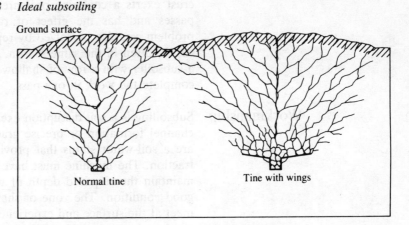

Ground surface

Normal tine

Tine with wings

Figure 12.4   Poor subsoiling

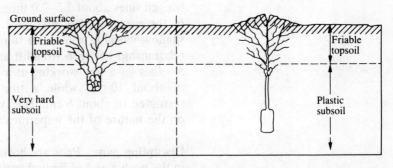

Ground surface

Friable topsoil

Very hard subsoil

Friable topsoil

Plastic subsoil

textured soils must not be unduly wet, while fine-textured soils must be fairly dry to be friable. Thus fine-textured soils are most difficult to subsoil successfully. Soils dry out from the surface downwards and the limit of penetration of the dry, friable zone at any time is called the **critical depth**. Subsoiling down to the critical depth will be successful. Below the critical depth it will be unsuccessful. The soil must be allowed to dry out until the critical depth has reached the desired level in the profile. Unfortunately this is rarely practical and it is usually necessary, to subsoil to a limited depth on the first occasion and to advance into the profile progressively as soil drainage improves in subsequent dry seasons.

To add to the difficulty, clay soils, when dry, become very hard and still resist shattering. Furthermore, the presence of a resistant crust exerts a confining pressure on the soil below as the tine passes and has the effect of raising the critical depth. This problem can also be solved by repeated subsoilings at increasing depth, often in the same season. If sufficient power is available, a subsoiler with a row of shallower leading tines may be able to complete the work in one pass.

Workmanship

Subsoiling is a soil disruption exercise and is not concerned with channel formation or precise gradients. The basic requirements are a soil-water status that provides friability and good surface traction. The machine must have sufficient power and weight to maintain the required depth of work and the shoes must be in good condition. The zone of shattering between tine runs must meet at the surface and experience will show the best tine or run spacings for a particular site. As a general rule, for normal tines the best spacing is about 1.0–1.5 times the working depth and for winged tines about 1.5–2.0 times working depth. The angle of lift of the wings is also important. A large lift of 10 cm or so for shallow working will cause too much surface unheaval. A good relationship between wing lift and tine depth is about a factor of 5 – that is, a tine working at a depth of 50 cm needs a wing lift of about 10 cm, while a tine at 30 cm must have wing lift restricted to about 6 cm. The type of subsoiling needed depends on the nature of the impermeability.

Disrupting pans   Pans are horizons of induration or compaction in the profile and of limited vertical extent, usually not more than about 20 cm. The measures taken to disrupt them depend on their depth within the profile.

Shallow pans   Soil capping and topsoil compaction caused by field traffic are broken up by normal seed-bed preparation activity. Pans in grassland or capping at the end of the wet season in fields cropped with winter-sown cereals are more difficult because the surface must not be unduly disturbed. A good method is to use a side-slanted tine cultivator which slices the soil at an angle of about 45° to the surface and penetrates to about 30 cm. The slanting blades lift each slice of soil gently to form a regular series of fissures.

Plough pans occur just below the topsoil and may be broken up by an occasional deeper ploughing or by attaching to each plough body a short, stout subsoiling tine, usually about 10 cm long. Multitine subsoilers with intermediate tine length and fitted with short, low lift wings are available for deeper cultivation and pan disruption. They work at any depth to a maximum of about 30 cm and remove the effects of panning quickly and cheaply.

Deeper pans Normal subsoilers are needed to disrupt natural pans which are found deeper in the profile. The depth of work required depends on the location of the pan and for best results the shoe needs to be set at least 10 cm below the indurated or compacted layer. Care is needed with natural pans which may vary in depth across the field. Most pans are easily broken up by a single pass of the subsoiler provided that the normal rules of good subsoiling are observed.

The need for underdrainage in profiles with a pan When subsoiling is required to break up a pan the need for drains in the field is really independent of the subsoiling operation. The decision should be based on the likely drainage characteristics of the site after the pan has been removed. On sites with a coarse-textured soil not affected by a high groundwater table it is likely that no underdrainage will be required. For finer textures and higher groundwater tables underdrainage will be required and must be installed before subsoiling is attempted. Drain spacings and the need for permeable fill must be based on textural classification.

Improving indurated profiles

Many soils are underlain by drift formations which are naturally indurated. This is particularly common in glacial till in which the induration is found throughout the entire formation and is made worse by stones which are virtually cemented into position. Only the topsoil has become friable because of the natural agencies of soil formation but it is often waterlogged because of the impermeable material beneath. The obvious remedy is subsoiling to sufficient depth but the work is difficult since there is no practical lower limit to the induration to allow a lifting action by the subsoiler shoe. The requirements are the heaviest, most powerful equipment available and a dry surface to provide traction. In most cases the indurated subsoil is dry in any weather conditions. In the hardest profiles wings should not be fitted and the work should not be attempted at a single pass. The depth of work should be increased at successive passes by 6–10 cm or so either in a single season, if time permits, or progressively over the years. The final task is usually to cart off large stones brought up by the subsoiler.

The need for underdrainage in indurated profiles   Where the profile is totally indurated there is no ready escape route for the perched watertable even when subsoiling is completed and any attempt to subsoil without underdrainage does nothing more than deepen the layer of soil that can be saturated, allowing machines to sink even deeper. Efficient underdrainage must always be installed first and the subsoiling must cross the drain lines at an angle that provides a reasonable gradient for the subsoiling runs. The drain spacing and the need for permeable fill depend on the likely permeability value of the disturbed profile. Many of the indurated drifts have quite coarse textures when disrupted and loosened so that drain spacings of up to 40 m can be considered.

Improving medium-textured subsoils

The finer grained non-clay soils – silts and very fine sands – are difficult to drain because they have a low permeability value when in a massive condition and the unstable nature of the soil particles causes structure to deteriorate easily when the profile becomes wet. Very often the only means of drainage is to lay field drains at 10 m intervals and to use permeable fill. Subsoiling will improve the profile permeability if it is done when the soil is dry but the work will need to be repeated at short intervals, perhaps every year for arable cropping.

Improving fine-textured subsoils

Mole drainage over collector drains with permeable fill has been found to be the ideal drainage solution for most clay soils. But many sites have produced disappointing results after moling because the channels have collapsed after only one wet season. When this happens there is a tendency to return to a closely-set underdrainage system or to seek improvement by using a subsoiler. **Underdrainage alone will not be successful**. Furthermore, the difficulties of subsoiling will be at least as formidable as those for moling. Such profiles are rarely indurated – the slurried drift has not been able to lose moisture – and the presence of a pan in the subsoil is of little significance in an impermeable clay. Thus subsoiling must wait until the profile dries to sufficient depth and this may occur infrequently since it is probable that a wet climate is at least in part the cause of poor soil stability. To have any chance of success underdrainage must be installed in the best possible conditions and the trenches must have enough permeable fill to connect the topsoil directly to the drains. Subsoiling work must then be restricted to the driest possible season and the depth of work must not exceed the critical depth. With good soil management and patience it will be possible to increase the depth of work progressively as improving soil structure allows the profile to dry to greater depth. These soils are always difficult to drain whatever method is used but, nearly always, the balance of advantage lies with mole drainage. Mole channels permit initial dewatering deep in the profile and repeat molings will improve soil stability more rapidly.

Permeable fill

The material chosen to maintain permanent permeability in the drain trench is called permeable fill. There is a tradition in many areas to use locally available cheap materials such as peat litter, straw or brushwood for this purpose and they are undoubtedly better than impermeable soil, but since they all can decay the permeability must be eventually lost. For this reason the term permeable fill is restricted here to mean material which resists decay, slaking or collapse in the drain trench. Such materials include gravel, broken rock, clinker and sintered fly ash. Technically, permeable fill would probably benefit most schemes. It would certainly do no harm, but the cost is high and it must be used prudently. However, in the many soils which can have trenchline permeability problems it is an expense that cannot be avoided.

The function of permeable fill

Permeable fill can be used for a number of purposes. It can be used as a **filter**, it can improve the hydrologic properties of a pipe drain or it can be used as **connector**.

Filters are used in drainage works to reduce siltation. Permeable fill is not a very satisfactory filter because the grain size selected must have pores which match the size of soil particle to be filtered and when the match is achieved the filtered material blocks the pores, rendering the whole assemblage impermeable.

An envelope of gravel increases the effective outside diameter of a pipe drain and can reduce concentration losses. But improvement is small and a true envelope is difficult to install. Such a function for permeable fill is rarely justified.

The most effective use of permeable fill is that of a connector. It provides an easy pathway to underdrainage pipes for water on the surface, water in subsoiled horizons, in mole channels and in old drainage systems being replaced.

Design standards for permeable full

The material used should be permeable in the trench but it must also flow easily from a hopper on the draining machine into the trench. Material which bridges in the hopper or chute will disrupt the flow leaving the drain without permeable fill in places. A size range of 5 mm to 5 cm is satisfactory and it is better if there is some uniformity of particle size. Unwashed sand and gravel is permeable but the finer grains can wash down into the drain pipe and cause siltation. The permeable fill must not be mixed with soil when stored for use and must be placed directly on top of the pipe. It should reach the required level in the trench and be evenly distributed along the trench. In slowly permeable soils it is essential that the permeable fill is brought to at least 15 cm above the mole channels or into the subsoiled layer. In less stable soils it is better to bring it up to be in contact with the more permeable topsoil. For connecting old drains it needs to span the depth separation of the two underdrainage systems. In all cases allowance must be made for settlement in the trench which may lower the top level of permeable fill by as much as 15 cm. To reduce the cost of permeable fill the drains should be installed at the minimum permissible depth and in the narrowest possible trench. Trenchless machines are best in this respect because they form a narrow slot but this may be offset by a decreased efficiency in the permeable fill due to trench-wall smearing by the passage of the blade and possibly by invasion of the voids in the permeable fill by slurried soil when field traffic occurs in wet weather.

When to use permeable fill

The following circumstances outline the essential use of permeable fill in a drain trench for the successful completion of the work:

1 Collector drains associated with mole drainage. Water must be cleared from mole channels as quickly as possible.
2 Collector drains associated with subsoiling if spoil backfilled in the drain trench is likely to be sufficiently impermeable to cause saturation of the subsoiled zone in the profile. This may cause slaking and loss of soil fractures, particularly in profiles with a high content of clay, silt or fine sand.

3 French drains set to intercept surface flow or interflow.

4 Underdrainage to be installed by trenchless drainers in all cases where an existing underdrainage system must be connected to the new pipes unless the soil profile is very permeable.

5 Interceptor drains that are set to collect artesian seepage but local gradients prevent drains from being located in the aquifer. They must be connected to the aquifer by permeable fill (see Ch. 15).

The following are sites where permeable fill is desirable but not necessarily essential:

1 Underdrainage in moderately permeable soil with or without subsoiling. The decision depends on the balance of choice between the cost of permeable fill in closely spaced drains and crop yield losses that may result from slow clearance of excess soil water. A compromise may be possible by placing permeable fill in every second or third lateral drain.

2 Interceptor drains set to control gravity-spring seepage. In some cases the interception effect is increased by using permeable fill.

Permeable profiles composed of sand, coarse sand, loamy sand, loamy coarse sand, sandy loam or coarse sandy loam are unlikely to need permeable fill.

Chapter 13 **Low-cost drainage**

Difficult sites for land drainage

Many sites that have soil drainage problems are not worth the cost of comprehensive underdrainage because the returns expected from the investment are likely to be severely limited or because there will be very high installation costs. Many areas are limited by difficult weather conditions which sufficiently inhibit plant growth to preclude all except the hardiest species. In sites like this, even where good soil drainage is technically possible, it will not be attempted for economic reasons. Similarly, in areas where more productive farming activities could be expected there are sites which are not comprehensively drained because of high costs of installation, especially if some major engineering works are required which cannot be justified or organised. There are also sites which cannot be drained satisfactorily however much is spent on them. These are usually unstable clays, silts or very fine sands located in areas of cool wet climate where the profile just never dries out. However, nearly all sites can justify some limited improvement measure as long as the economic benefits are carefully evaluated.

Low-value land
Site classification

Land may be of limited value for agriculture because it is situated at high altitude or latitude, in the path of cold winds, in a high rainfall area or subjected to a combination of such factors. Limitations of this type are typical of the exposed margins of the maritime climatic regions. Agricultural land use is restricted to extensive livestock rearing on unimproved rough grazings, most of which are unenclosed. The rough grazings consist of natural or semi-natural plant communities and are variously described as **moorland**, **fell**, or **hill grazings**. Associated with the moorlands, in the more sheltered valleys and nearby better lowlands, are enclosed fields which have improved grassland and sometimes other crops. These often are the 'inbye' fields of hill farms and are used to produce winter fodder and provide more sheltered pastures for the livestock. This type of land has greater potential than the open moorland but may still have limitations in terms of land-drainage expenditure. These somewhat less restricted sites are classified as **marginal land**.

Drainage improvement in moorlands

Moorlands include a wide range of plant habitats from high mountain slopes to coastal lowlands. Soils are either organic or leached mineral types which are often very shallow or stony. The greatest potential for drainage improvement occurs on more level sites in valleys or on plateau areas where the soils are deeper. Drainage work is limited to clearing the surface water by cutting small open channels and beneficial effects take place gradually as the plant habitat changes. In time the grazings may improve and the land may support more livestock.

Hill drains

The moorlands cannot justify the use of normal types of drainage machines including mechanical diggers. The only practical drainage improvement is the cutting of a system of small surface channels which are described as **hill drains** or **moorland grips**. The drains are ploughed out as a continuous furrow leaving neatly-formed channels about 40 cm deep with a half-rounded bed of about 25 cm diameter and a channel top width of about 70 cm. The actual dimensions are not critical as long as the side walls are stable and the channel is at least 30 cm deep. It is essential that the ploughed-out furrow is set down on the down-slope side of the channel at least 55 cm back from the edge. These ideal dimensions are easily obtained in peat or soft mineral soils but the depth of cut may be restricted in thin mineral or in stony soils.

Siting hill drains

Hill drains occasionally may have individual outlets to some suitable existing channel. More often a system of leader and lateral channels is required. The leaders are opened up by the same machine but set slightly deeper to provide an outlet and the junctions should be formed as neatly as possible. It is often worthwhile to clean up the junctions using a spade. Outlet points for the leaders should be located in natural channels or other hollows from which the drainage water can escape with the leaders following the line of natural hollows wherever possible. The laterals should be cut across the major slope to have the best possible interception value for surface flow and should be aligned in such a way that the channel has an even gradient to permit continuous flow all the way to the outlet. The flow velocity during peak flow rates should achieve some self-cleansing of the channel without any bed erosion. The drainage pattern may be herring-bone, along the contours or random in layout depending on the nature of the site. Lateral channels should not exceed about 400 m in length and should not have very sharp bends if channel erosion and subsequent deposition and blockage are to be avoided. Lateral channel intervals of about 20 m are satisfactory but the spacing is not critical and may be wider for reasons of economy or surface topography. Great care is always needed to avoid causing gully erosion which might scar the hillsides. This applies mainly to the leader channels which must flow more in line with the major slope to collect water from the laterals, particularly where they occur near the steeper margin of a plateau area. Where possible, existing natural channels should be used as the main outlets and it is important to select a layout that permits the best possible gradient control. In any event channels should never be cut down a steep slope and large hill drainage systems should never discharge down a steep slope that has no existing natural channel.

Machines for hill drainage

There are two basic types of machine suitable for opening hill drains – large drainage ploughs and rotary ditchers. For the extensive areas of rough surface found in the moorlands only the drainage plough has the necessary strength and speed of operation to do the drainage economically. The drainage plough is a modified forestry plough which cuts out a furrow of the required dimensions and places it to one side, leaving a neatly formed channel. No gradient control is possible and the plough is simply pulled across the ground in a direction which provides a suitable gradient. The huge plough body is attached to a heavy frame carried on two very large rubber-tyred wheels which keep the plough at the selected depth of work on, sometimes, very soft surfaces. The whole implement achieves some gradient smoothing over minor undulations. To permit economic rates of work the plough must be towed by a powerful crawler tractor fitted with wide, low ground-pressure tracks and a winch is necessary to retrieve the plough when it sinks into a hidden mire. A quick-release mechanism is needed to protect the plough from damage when it hits a boulder. The small rotary ditchers consist of vertically, or near vertically, set cutting wheels which spin out a small surface channel. They are attached to the linkage of farm tractors and driven by the power take-off shaft. They have a slow work rate, are easily damaged by stones and have poor cross-country mobility in the worst conditions. However, they are cheaper to buy or hire and can be very useful for the smaller operations in marginal and arable areas.

Maintenance work

Hill drains deteriorate year by year as a result of erosion, deposition or bank collapse resulting from flooding, frost or the trampling of cattle. It is usually necessary to repeat the channel-cutting operation every 10 years or so if the improvement is to be maintained. Where economic factors make this difficult the working life of the scheme may be extended by improving the leaders only and reforming the channel junctions. It may also be worthwhile clearing any localised points of bank collapse along the lateral channels using hand tools where easily-cleared ponding is identified.

Consultations

There are some non-agricultural matters to be taken into account when operating in the moorlands. These extensive, sparsely populated areas often form the catchment of a water supply system for a major urban settlement and any surface drainage work may be controlled by local or national legislation. It is essential to discuss the drainage work with the water supply authority before any work begins. These remote areas may also be of some scientific interest. They may form part of a nature reserve or contain some remnants of ancient earthworks or other features of historical interest. Many areas are protected by legislation or are otherwise of great concern to the public so that it is always advisable to consult the relevant authority where any surface disturbance is considered.

The marginal areas
General features

Marginal lands include a range of field conditions from quite reasonable cropping land to recently enclosed parts of the moorland improved by shallow cultivators or other surface treatments. Natural profile drainage tends to be slow and difficult in these areas of usually cool moist climate and all except the most freely draining soils on water spreading sites are likely to lie wet through most of the year. Furthermore, the land usually lies close to large elevated catchments and steep slopes so that the lowest areas, at least, are subjected to severe flooding from time to time. It is essential that the main natural drainage channels and ditches should be of adequate capacity to clear the flood water in a reasonable time and that the channels be kept in good condition. Unless there is an exceptionally close network of ditches it is unlikely that ditch piping on a large scale will be either possible or desirable. Certainly all interceptor and carrier ditches should be left as open channels. Since natural processes will have created a close system of drainage channels it is unlikely that large arterial works will be needed to create new channels although considerable improvement of existing ones may be needed from time to time.

Land drainage

A wide range of land drainage measures may be considered for these areas. Where the land quality and financial resources justify comprehensive underdrainage, the normal design factors and installation standards apply. In the least worthwhile areas – difficult peaty hollows or partly improved areas of nearby moorland – only hill drainage may be possible. The hill drains should be opened up just as described for moorlands. In these marginal lands it is usually a small-scale project where a rotary ditcher is quite satisfactory. Between the best land and poorest inbye land there are many drainage problem areas that can best be tackled by some intermediate type of drainage technique. The method chosen depends on the nature of the drainage problem and the resources available.

Springs

A common feature of the marginal land areas is the irregular pattern of the wet patches resulting from various types of spring seepage. Since the installation of interceptor drains to control spring water is a relatively cheap drainage operation, underdrainage will be the obvious choice in nearly all cases. However, if the springs affect a wide area and the land is poor or otherwise difficult to drain because of a steep surface or shallow soil, it may be necessary to restrict the drainage improvement to the cutting of an interceptor ditch just below the springline, near each group of springs or at the bottom of the slope as appropriate. This will prevent the spring water from spreading over a large area of land.

Surface flows Inbye fields often lie at the bottom of hill slopes and can receive considerable inflows of surface water from nearby moorland. Surface inflows can always be cured relatively cheaply by cutting interceptor ditches along the upper margin of the area to be protected.

Intermediate underdrainage Large areas of marginal land are wet for considerable periods each year because of the cool wet climate. Where the soil is also slowly permeable, comprehensive drainage can be difficult and expensive and will be justified mainly in the better fields often close to the steading. For the other land it is worthwhile to adopt some intermediate drainage system since open hill drains will not be desirable. A number of options are available:

(a) Underdrainage with permeable fill up to the level of the base of the topsoil and aligned across the main slope may be set out across the land in a regular pattern but at wide intervals (perhaps 80 m or more) as determined by available funds.

(b) Where soils and gradients are suitable the land may be moled as an independent operation – with outlets into a ditch – at spacings and remoling intervals that are determined by the weather conditions and economic factors.

(c) Gravelled moles may be installed. Mole ploughs fitted with a hopper and chute create mole channels and fill them with gravel to prolong the life of the channel.

(d) Gravelled slots may be installed. Narrow slots about 10 cm wide may be opened up across the land using some type of narrow rotary ditcher and the slots filled to near the surface with gravel.

The essential value of intermediate drainage is the rapid removal of surface water after rain and the chosen system must have a good connection to the soil surface. Gravelled moling gives the channels a longer life in difficult soils but it is very expensive when the moles are set out at normal spacings. When set at wider spacings the drainage effect is even more dependent on connection to the surface which may be easily lost by surface compaction in the very soils where the technique is most likely to be used. Permeable fill can be very expensive unless there is a nearby source on the farm and in many cases little is saved by not using pipes. Perhaps the best course of action is to lay field drains, with permeable fill, initially at wide spacings. This permits future intensification, and also soil treatment, all subsequently leading to comprehensive drainage as time, conditions and funds permit.

Difficult drainage in more fertile areas

Even in the more favoured farming areas there are sites that are difficult to drain. The problem may be associated with the type of soil or with some topographical or other restriction of the site.

Soils that are difficult to drain

There are some soil types that can cause special problems for the land improver. They include soils which can be unstable – and these are particularly difficult on the wetter sites – and soils which cause the precipitation of solids from the drained soil water. The difficult soils include:

- Profiles composed mainly of silt or very fine sand that cause persistent and rapid siltation.
- Unstable clay profiles which become liquid when wet and tend to flow into pipe slots or to invade pore spaces in permeable fill. The clay may dry out in this position totally sealing off the pipe.
- Peat which can be very unstable and in a similar manner can block pipe entry slots. Attempts to dewater peat can be an uncertain process in a wet climate.
- Ochreous soils which can block pipes. In most cases some underdrainage solution is available as discussed in chapter 15. In the worst conditions underdrainage may not be practical.

Drainage options available

If the climate is persistently wet, few soils are worth draining. Even in moderately wet areas clay profiles or peat soils may dry out only rarely to permit improvement. Difficult, unstable sandy or silty soils and ochreous soils are more widely encountered. In most cases there will be a history of drainage failures in the districts concerned. As with marginal land, decisions on the amount to be spent on drainage depend on the costs and expected returns. The options available are hill drains (hill gripping) or land drainage ditches with or without surface grading between ditches. Surface grading is an effective option for land with a reasonable potential but the work can be expensive and it is an engineering operation. It is often necessary to strip and store the topsoil beforehand so that it is not lost during regrading work.

Sites that are expensive to drain comprehensively

Where a site cannot be drained comprehensively without resort to pump drainage and this is unlikely to be cost effective it is necessary to consider other options. In this type of site well-maintained, relatively close ditches are essential to keep the ground-water table as low as possible. Further work then depends on the gradients available and may be limited to hill drainage channels or some intermediate drainage option.

Chapter 14 **Pumped drainage**

Many sites do not have an outlet for drainage by gravity flow. In the majority of these the only feasible method of land improvement is to pump the water into an outflow channel that is elevated above the level of field drains.

Low-lying sites

On many sites set in coastal lowlands, near inland waterways or in isolated upland hollows the general land surface may be at or near the local base level. In some cases a partial improvement may be obtained by a system of flat-gradient ditches depending on the relative levels of land surface and outlet channel. No improvement is possible where the outlet water level is above the land surface. In most natural settings this implies a surface covered by water but where land has been reclaimed from inundation the site will be protected by flood banks. In the long settled, intensively cultivated lands of the major river valleys and deltas the land surface may have sunk gradually below the local river level in response to soil shrinkage in the fields and accretion in the river channel. Successive generations of farmers will have increased the height of the banks. Wherever land lies behind a flood protection barrier the land-drainage problems are increased but there are many variations. The outlet channel level may restrict gravity drainage permanently, seasonally in response to climatic change or daily near tidal waters. Where land-drainage improvement is required the only available method is to lift the drainage water into the outlet channel by means of water pumps or other water-lifting devices.

The size of the project

Pumped drainage can be technically feasible in a wide range of sites from large regional projects like polder reclamation in the

Netherlands to very small schemes serving only one or a few lines of field drains. All the larger operations require the organisation and funding of some central authority while many lesser projects depend on the co-operation of a number of land owners. All work at this scale of operation is relatively complex and must be controlled by qualified engineers. However, small projects, involving only one farming business can be straightforward and within the scope of an individual land improver if the principles are understood and specialist advice is obtained where necessary. The first task is an assessment of project feasibility.

Project feasibility

Any pumped drainage project must be both technically feasible and economically worthwhile. In most cases the economic limitations are more severe but there are sites with major physical limitations. Furthermore, the two factors are not independent since physical difficulties increase the cost. At any scale of operation pumped drainage requires greater capital investment at the installation stage than does the equivalent area of gravity drainage while the annual cost of running the pump must also be supported by extra output from the drained land. Generally, it is only the best sites with a potential for high productivity that can justify the expenditure by growing horticultural crops, high-value arable cash crops or fruit orchards. Intensive dairy enterprises may also justify an extension of the productive grassland, particularly where the watertable can be controlled to permit continued growth of grass during a dry season. Elsewhere small projects may be justified if they permit some improvement of the general farm layout, improve access to a greater area of land and generally aid management in addition to increasing the available area of land. The gains should be above average for under-drainage improvement and there should be a minimum of physical limitations associated with the site.

Physical limitations

The nature and extent of any physical limitations present and the work load to be placed on the water-elevating mechanism must be evaluated by survey of the site to be drained and the surrounding area. This will include a reconnaissance of the whole catchment, examination of the soil properties, some assessment of the underlying strata and a careful measurement of all gradients and levels within the site. The physical limitations can be arranged in four broad groupings.

1 The volume of water that must be pumped out of the area.
2 The height that the pumped water must be elevated.
3 The general suitability of the site for the installation of pumping equipment.
4 The effect of drainage on the local environment.

The amount of water to be removed from the site

A factor of high priority in evaluating a pumped drainage project is an assessment of the amount of water that must be removed from the site each day to maintain adequate soil drainage. Like all other drainage work, this depends on the climate, the extent of surface inflows and any groundwater inflows through aquifers. In low-lying sites there can be no significant surface outflows; outwards groundflow is very unusual.

Rainfall Incident rainfall should not be a serious problem in potentially productive areas. The design rainfall rate can be selected in the normal way for the intended cropping programme, drainage type and site characteristics. Usually five-day rainfall periods will be selected and the pump chosen should have the capacity to clear the five-day rainfall amounts in five days for rainfall of the selected return frequency values.

Evapotranspiration Although pump capacity is based on winter season rainfall events in a maritime region, the number of days each year when the pumps must operate is determined by the balance between rainfall and evapotranspiration. This is important for running costs. For areas with similar annual rainfall the cost of pumping increases as the evapotranspiration rate decreases.

Surface water inflows All sites requiring pumped drainage are likely to be collecting points for overland surface flows. Inflows will be least where the site is part of a wide area of flat lowlands and greatest where the site is an isolated hollow within a more elevated catchment. Surface permeabilities within the catchment will also influence the amount of inflow. The extent to which surface inflows can be reduced will have a considerable influence on the cost of pumping the drainage water.

Groundwater inflows Inflowing groundwater is more difficult to recognise and to quantify. Generally, it is more likely to be a problem for any drainage sites set in permeable catchments where sandy soils directly overlie aquifers. Attempts to pump-drain a small part of a flat permeable catchment will cause lateral ground-flow into the drained sites in response to the lowered groundwater table. In such a site the land can be drained only as part of a communal scheme to drain the whole catchment. Groundflow of this type will be negligible in less permeable catchments.

Channels cut through permeable soils permit leakage into the soil nearby when the water level in the channel is higher than the local groundwater table. This may cause the recirculation of pumped water. Where leakages are not great the project may still be worth while but appreciable return flows increase costs beyond economic limits. River flood plains in glaciated landscapes often have so much return flow that pumping is ineffective.

The distance that the pumped water must be elevated

The basic features of a small-scale pumped-drainage scheme consist of a normal underdrainage system discharging into collector drains or into collector ditches which carry the water to a pump **reservoir**. The reservoir holds the water long enough to collect a sufficient quantity to give the pump a reasonable running period and keeps the **sump** beneath the pumping mechanism filled with water. Electrical switching gear can be installed to control the pump automatically as required by the rise and fall of water in the sump. When operating, the pump expels water from the sump through pipes, upwards towards the outlet channel. The outlet end of the pipe is fitted with a non-return flap-valve to prevent any backflow. Thus the sump is the effective gravity outlet point for the underdrainage system, with its depth being determined by the total requirements of sufficient cover for lateral drains, drain gradients, ditch gradients and reservoir gradient. The differences in level between the surfaces of water in the sump and in the outlet channel is the distance that water must be elevated, which is called the **static head** of the system.

The general suitability of the site for pumped drainage

Several site characteristics can determine whether or not a piece of land can be drained. In most cases the problems can be solved technically but the cost of doing so may be prohibitive. The following site features should be considered.

Site flooding   Some types of site, like small depressions in a large catchment area, may be so regularly and seriously flooded that either drainage is not worthwhile or the pumping control gear can be damaged by the high flood levels.

Sound foundations   Pump mechanisms are usually housed inside a substantial pump house or chamber constructed from normal building materials and must be stable enough to be free from tilting, cracking or sinking. For this purpose a firm foundation is essential. Low-lying drainage sites are naturally very wet, may well have a cover of peat and are not ideal as sites for buildings. Decisions to proceed will depend on finding a suitable position for the pump house.

Accessibility   There should be easy access to the pump house for power supply, delivery of fuel, maintenance and supervision. It may also be necessary to be able to observe the pumphouse from the main buildings if vandalism is a local problem. All of these requirements can be met by siting the pump as near to a road and as near to the farm as possible. The cost of constructing a new road through unstable land may be prohibitive.

Soil stability   Soils composed of the less stable soil particle sizes are easily dislodged by the vigorous flows associated with pump drainage. In particular, the banks of the reservoir ditch may be difficult to stabilise. Also important is the high cost of maintenance required where the pump has to deal with a mixture of sand and water which quickly wears the pump mechanism.

Environmental constraints   There are non-agricultural factors to be taken into account. Poorly drained marshlands and peatlands often form unique and isolated natural habitats for wildlife that are easily destroyed by drainage or even attempted drainage. Any alteration of the site environment will be of concern to the general public and the particular habitat may well be protected by legislation. Similarly, there may be amenity interests in the area. In all cases it is advisable to ensure that the project is acceptable to all relevant authorities before work begins.

Assessing project feasibility

The decision to proceed depends on balancing the costs of all the identified capital investment needs as well as the likely pumping costs against the estimated returns. Improvements made to reduce pumping costs may well increase the capital costs and these also must be taken into account.

Preliminary site improvements

Once the decision is made to instal pump drainage the next step is to carry out the improvements needed to reduce, as much as possible, the amount of water that must be pumped. This requires isolating the pump-drained area from all types of inflows as far as is technically possible and by making maximum use of any available gravity outflow.

Gravity flow

Drainage by gravity flow should be used wherever it is available. This is of value in several ways.

On marginal pumping sites   On sites which can be partially improved by gravity flow, the groundwater table can be lowered to an appreciable extent by cutting new ditches or improving existing ditches to the best possible standard. The pump is then needed only to complete the work of lowering the groundwater table. Existing ditches are very often in poor condition since their improvement without the assistance of pump drainage has been seen to be futile. On this type of site it is better to keep the pump reservoir and outlet channels separated from gravity flow channels to avoid backflows in time of flood.

Where the water level in the outlet channel fluctuates   Wherever the level of water in the outlet channels falls regularly or seasonally to a level that permits gravity flow drainage this should be used by installing a gravity outlet via a sluice through the bank and fitted with a good quality non-return flap valve.

Flood water   All sites should have provision for gravity-flow escape of flood water which otherwise would be retained behind the protective banks at a level above that of the water in the outlet channel.

Isolating the site from inflows

Appropriate action depends on the site. If the site is small and part of a small catchment that does not generate significant surface flows it is likely that efficient underdrainage in the higher areas will eliminate surface flow into the site requiring pumped drainage. On any site drainage of surrounding areas will have some effect in reducing surface inflows.

Where the drainage site is part of a wide area of flat land at about the same level it is generally found that the whole area is served by a close network of flat gradient ditches. In most cases it is possible to isolate the site simply by blocking off the inflow ditches. The blocked ditches can then drain via the rest of the network and bypass the drainage site.

Where the drainage site lies below the level of the rest of a catchment large enough to generate appreciable surface flow the site should be protected by constructing interceptor ditches along the base of the slopes surrounding the area to be drained. The interceptor ditches must discharge by gravity flow outside the site and where awkward hollows occur it may be necessary to construct embankments to continue the line of the channel.

Where the drainage site lies distinctly lower than the rest of a catchment large enough to generate a stream it becomes necessary to divert or elevate any stream or carrier ditch that enters the drainage site. The channels must be contained by banks on one or both sides depending on whether the route crosses the low area or passes along its margin. The elevated channel may often be used as the outlet channel for the pumped drainage. Many sites of this type will end up completely surrounded by protective banks.

Siting the pump

Selecting a suitable place for the pump is a difficult decision involving a compromise solution to meet some contradictory requirements. No two sites are alike and general advice is of little value. The choice must allow for four basic requirements:

1 The pumphouse must be sited on a firm foundation.
2 There must be easy access to the pump site.
3 The pump must be located as near as possible to the point which forms the best gravity outlet for the drainage system, i.e. the lowest part of the drainage site. Any other location involves longer and deeper collector drains or ditches and increases the depth needed for the sump.
4 The load on the pump must be carefully controlled. For this

purpose it must be located close to the outlet channel so that the delivery pipes are as short as possible to reduce friction and other losses of efficiency. Where the pump cannot be placed near the outlet channel it is necessary to construct an extension of the high-level channel to reach the pump site.

The pump workload

For practical purposes the workload on a land-drainage pump is the amount of water that must be elevated to a given height in a given time. The volume of water is normally the pump catchment area multiplied by the chosen design drainage rate. Allowance may be needed for return groundflow which can recirculate pump water from the outlet channel. The elevation required is the static head of the system but allowance must be made for **head loss** through the pump caused by constrictions, friction and turbulence. The static head and head loss together constitute the **total head** and represent the work performed by the pump. The head loss is proportional to the flow velocity through the pump and, for small schemes, usually with spare pump capacity, it need not be a significant factor.

The types of water pumps available

Water-lifting devices of various types have been in use since ancient times and they can still be seen, little changed, in the arid lands. They are usually operated manually or driven by harnessed animals or occasionally by windpower. Modern derivatives or new designs powered by electric motors or internal combustion engines are currently available for land drainage. Each type has a characteristic performance and suits particular pumping requirements.

The Archimedean screw

The open **Archimedean screw** consists of an inclined shaft on which is mounted two or three blades in the form of a helical spiral which lies in and fits neatly a semi-circular chute. The lower end lies below the level of the water in the reservoir and when the spiral rotates the water is trapped between successive blades and the chute and is carried upwards to the top of the chute from which it cascades into the outlet channel. This simple device with low running speeds is cheap to run and easy to maintain and can cope with weeds, debris or sediment-laden water. Choice of spiral length and diameter as well as running speed permit its use for a considerable range of discharge values but its static-head capacity is limited to about 5 m for a single-stage unit. The major disadvantage is that the open mechanism cannot be fitted with a non-return valve and any serious rise of water level in the outlet channel can cause flooding through the installation. For this reason it is best suited to sites with fairly static water levels.

Impeller pumps

Modern water pumps consist essentially of an **impeller** of two or more blades mounted on a shaft inside a casing. Rapid rotation of the impeller forces water to flow inside the casing and the pressure developed pushes the water through the piping to a higher level. There are three basic types of impeller pump depending on the action of the impeller.

1 The **axial flow** pump has the simplest mechanism. The rotating impeller causes water to move along in the direction of the drive shaft or axis of the pump.
2 The **centrifugal flow** pump has a different action. Rotation of the impeller causes water to be thrown off radially and tangentially into a bulbous collecting casing.
3 The **mixed-flow** pump is an intermediate type combining the effects of axial and centrifugal flows.

Depending on the size of the pump and the running speed of the shaft, each type is capable of dealing with a wide range of discharge values.

Selecting the most suitable pump

For larger scale pumping operations, selection of the most suitable type is based on differences in the characteristic efficiency curves of each design. In general terms centrifugal pumps are best suited to the greater head values (usually those more than 10 m); axial flow pumps and Archimedean screws are best suited to smaller head values while, as might be expected, mixed flow pumps are best for intermediate values. For small scale schemes axial flow pumps are usually chosen and are simple to instal. A range of efficient, self-contained pump-and-motor units of small capacity is available for the smallest discharges. These may be either axial-flow or centrifugal-flow types. For known static head and discharge values the pump size can be selected from the specifications of a range of pump sizes as supplied by manufacturers. Each will show the most efficient size for different work-loads.

There will be a choice between a larger pump running at low speeds or a smaller pump running at faster speeds.

It is a matter of balancing initial costs against maintenance costs. A small margin of error is acceptable as long as the pump has sufficient capacity to clear peak design flow rates but it must not have such excessive capacity that it evacuates the sump faster than the reservoir can supply water. This results in a rapid series of stops and starts, causing excessive wear. The pump chosen must be self-priming and the best power source is a three-phase electrical supply although single-phase is satisfactory for smaller motors. Self-starting internal combustion engine units are available for sites remote from mains electricity.

Controlling a pump

For farmer-installed pumped drainage systems there is no resident pumphouse attendant to observe water levels so pump operation must be automatic. A range of automatic control systems is available, activated by changes of water level in the sump. A simple but effective device is a float attached by a line to a counter-weighted toggle switch. This is an inexpensive system but it is easily damaged and the range of water levels in the sump is severely restricted. An unlimited range of water levels can be controlled by two low-voltage probes suspended on cables above the water in the sump. Water touching the higher probe activates the circuit to start the motor which then continues to run until the circuit is broken by the water level falling clear of the lower probe. This system is suitable for a wide range of schemes and is reliable except where weeds or changes of water conductivity interfere with the electrical circuits. Tilting mercury switches suspended on cables are perhaps best for the smaller project. They operate in a similar manner to probes; but the circuit is made when the upper suspended switch begins to float and tilts from a vertical to a horizontal position and is broken when the lower switch is above water and vertical. Water in the sump is not part of the circuit. They can be upset only by swirling water and weed growth. It is a simple matter to adjust the length of cable to the chosen water levels in the sump.

Protecting the pump

Various design factors, ancillary fixtures and attention to routine maintenance are required to ensure that the pump remains in good condition. All impeller pumps are badly damaged if allowed to run empty. This should not happen if the switching gear is well maintained and correctly set. The switch gear control mechanism must be well protected, kept dry, clear of weeds or debris and should not be required to switch on and off so frequently that relays and contacts wear out. Choice of correct pump capacity and the construction of a reservoir of suitable size should eliminate this problem. The switching circuitry should be protected either inside the pumphouse or in a watertight box set well above the highest possible flood water level. Impeller pumps can be damaged if they pull in weeds or floating debris and wear out rapidly if they pump a mixture of sand and water. To avoid all such damage the inflowing water must flow through a weed screen grid and silt-collecting box of the type described for open inlet drains in Chapter 11. Regular raking of the screen and removal of silt from the box are essential. For all except the smallest systems it is best to construct a sound screen-clearing platform to allow easy weed raking. Finally, pumps of the self-priming type will be damaged if the water freezes. All pumps sited where frosts can occur must be protected by sufficient insulation or kept above freezing point by heaters.

The choice of drainage system

All larger pumped drainage schemes and most smaller scale projects are based on a system of ditches and open reservoirs. The very flat gradients associated with pumped sites makes open channels an essential feature for all except the smallest areas. However, where the site extends to a maximum of about 10 ha and pipe gradients can be achieved, direct pumping of under-drainage water is feasible.

Open-channel systems

Most pump drainage schemes are based on open channels because of the need for large channel capacity and to allow for slight gradients so common on pumping sites. Very often all of the collector channels for the underdrainage systems are also open ditches. The important levels should be calculated by starting at the field underdrainage and working towards the sump. The field drains must have sufficient cover, gradient and an outlet 15 cm above water level in the collector ditch. The ditch system and reservoir must have gradient towards the sump. This determines the depth needed for the base of the sump. There may be a special reservoir area or the carrier ditch may act as a reservoir but it should be able to supply water to the pump as fast as the pump can evacuate water from the sump. Reservoirs are subject to frequent and rapid changes of water level and are prone to bank collapse. In all except the most stable bank-forming materials it will be necessary to support the banks by some means as described in Chapter 10.

Reservoir capacity

The selected rate of discharge needed to drain the area determines the pump capacity needed. The reservoir capacity must be sufficient to permit something like 1 hour of continuous pumping between switch-on and switch-off water levels in the sump. Where reservoir capacity is restricted for any reason the minimum satisfactory running period is 20 minutes. Thus the volume of water to be stored can be determined. One dimension of the reservoir is fixed by the base of the sump level and the maximum height of water in the ditches which leaves the field drain outlets clear. The other dimensions are then obtained by selecting sufficient length and width of channel. The only limitation is the need to provide a gradient towards the sump so that water flows in as quickly as the pump evacuates it. This requirement may restrict the length of the reservoir which must then be made wider to compensate.

Direct pumping of underdrainage systems

The essential features are similar to those of an open channel system except that all channels are piped. Because of limited capacity such schemes are limited in terms of the area of land that can be improved.

The reservoir is in the form of a large-diameter pipe protected by a silt trap and inspection chamber. The reservoir capacity is calculated exactly as for open reservoirs and the volume required is converted into pipe diameter and length to allow sufficient pump running time without causing land drains to be full of water. Usually a small combined pump-and-motor unit is set inside an inspection chamber below water level with the switching gear, suitably protected from the effects of weather, fixed to secure posts or set on a firm stand high enough to be clear of any flood-water. The cut-out level must be adjusted to the lowest level in the reservoir pipe that has sufficient discharge to keep the sump filled when the pump is running. With an all underground system there is no problem of weed growth or debris accumulation. Only the silt trap needs regular attention in addition to pump maintenance.

Layout for underdrainage   The sump and reservoir act as a gravity outlet for the underdrainage system. There are no unusual features in selecting a pipe layout other than those normally associated with the drainage of flat land.

Economic factors   The decision to invest in a pump system to drain land requires confidence in future market prospects. Small changes in fuel costs or market returns can completely alter the economic calculation. Nevertheless, provided the land is potentially fertile, the drainage is technically feasible at reasonable costs of construction and there are no environmental or legal restrictions, many projects can be worthwhile. For pump drainage installation it is worthwhile consulting someone with experience in such work to avoid costly mistakes. There are options for reducing running costs. A very economical system for all except the smallest scheme is to install two or more pumps in a pump well set to cut in at increasing sump water levels so that total pump capacity is variable and increases automatically to cope with peak discharge rates. Savings may also be possible by switching off the pump when the land is to be bare over winter or, for pasture, when grazing is not possible. Running costs should be carefully monitored. Pumps and ancillary equipment must be carefully and regularly maintained so that they run efficiently.

Chapter 15 **Drainage design**

Drainage layout Choice of layout depends on the identified drainage problem, the site topography and variations of soil present. Drain intervals in regular systems are a part of the layout although choice of spacing does not alter the basic pattern. Drain depth may influence spacing but usually does not alter layout unless depth restriction in the site makes redesign necessary. A system composed of a series of individual drains, each with direct outlet to an open channel, is called a **singular layout** while a system of laterals and leaders all discharging at a single outlet is called a **composite layout**. Larger areas of land may need additional outlet channels involving choice of sites for new ditches. Open channel drainage may be arranged in a variety of layouts similar to those for underdrainage.

Layout for ditches Where land is to be drained systematically by open channels only, the choice of layout and channel spacings can be based on the same criteria as for underdrainage. In some projects the open channels will be associated with land-grading work and in such cases the spacing is influenced by a complex set of engineering problems. However, for the majority of drainage works, ditches are intended to act in association with underdrainage. In such cases some different criteria are involved in siting ditches and decisions are needed at an early stage of the design process. Choice of route depends on a number of factors.

Function of the ditch Carrier ditches conduct water towards the main outlet point and should follow the line of any natural hollows or valleys, towards the lowest part of the site. Choosing other routes greatly increases depth of excavation and, therefore, costs.

Collector ditches should be set out as short and straight as possible along the lower margin of the area to be underdrained.

Intercepter ditches must be sited as required to control surface flow, usually along the upper margin of the site to be protected. In most cases a single ditch is sufficient to divert surface water originating outwith the improvement site. For large areas, however, there may be enough surface flow generated within the site to cause erosion. This occurs on sloping areas if surface water collects in sufficient volume to achieve erosive flow velocities over the surface, particularly on cropping land which can be bare at certain times each year. Depending on gradient, soil stability, soil permeability and rainfall intensity on the site, action must be taken to control erosion and this may take the form of parallel intercepter ditches across the slope. If there is no local knowledge of suitable ditch intervals for such installations it is necessary to consult a qualified engineer.

Water table control ditches, when used in conjunction with underdrainage, have the same function as collector ditches but are associated with flat sites. They are arranged as a parallel series or as a double parallel series dividing the land into rectangular parcels. The interval between ditches is based on the practical length of lateral drains and usually should not exceed 500 m.

Other factors

The routes of new ditches will be influenced by gradients, undulations, and rocky outcrops present on the site.

The route chosen should if possible avoid unstable soils associated with spring rises and other causes of bank instability.

Making due allowance for function and site characteristics, all ditches should be as straight as possible to provide the most efficient and easily maintained channel.

The requirements of underdrainage systems must be borne in mind. The practical limit for lateral drain length is about 400 m and on very flat sites it is better to restrict the length to 250 m.

Total length of ditches should be kept to a minimum that is compatible with drainage function. This minimum length per unit area increases as gradient decreases.

Where alternative routes are available the chosen location of ditches should be compatible with other aspects of land management, especially with regard to field size and shape.

Layout for underdrainage

Like ditches, the layout for pipe drains depends on their required function and site topography. Carriers drains work most effectively when they occupy valley bottoms and low points leading to the discharge channel, collector drains must lie along the lower margin of the site to be drained while interceptor drains need to be set out near the upper limit of wet areas. Drainage of spring sites is discussed separately but is concerned with siting interceptor drains. On the great majority of sites, lateral drains are laid to control water tables and are required to cover the whole site systematically. Thus layouts may take the form of regular patterns or some irregular system of siting. Depending on the drainage problem, the layout may need to cover the whole site or serve only parts of it.

Irregular layouts

Drain lines that are not set out as a regular pattern are needed in a number of situations. **Single** new collector drains may be needed to re-activate an old system of underdrainage. Undulating sites may be mainly dry but contain a number of small hollows with restricted drainage. Such sites need a **natural layout** with drains following the various depressions to reach the wet hollows. This is illustrated in Fig. 15.1

Sites affected by one or only a few widely separated but clearly defined wet patches (usually caused by springs) may be improved by the same number of **individual** drains tapping the spring water and conducting it to the nearest outlet point. On sites affected by

Figure 15.1   Natural layout

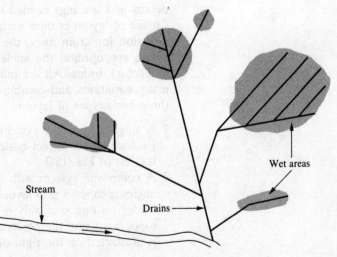

Wet areas

Stream

Drains

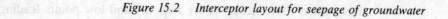

Figure 15.2   Interceptor layout for seepage of groundwater

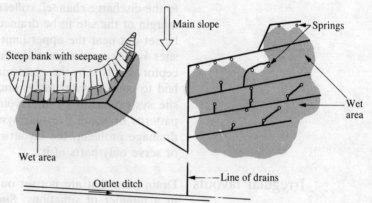

more closely grouped springs, by more diffuse seepage or by surface inflows, an **interceptor layout** is needed. The actual siting of drains depends on the pattern of wet patches and is usually irregular although parallel drains may be needed on larger spring-issue sites. Figure 15.2 depicts a typical interceptor layout for springline seepage from a steep bank on the left and drainage of a group of artesian springs on the right.

Regular layouts   Where systematic drainage of the whole site is required a regular layout is needed. This involves a series of parallel drains set at depths and spacings needed to cope with the drainage problem. Choice of layout is then a matter of selecting the most effective direction for drain lines, the arrangement of drain outlets and, where appropriate, the angle between lateral drains and leader (collector) drains. All are influenced by site gradients. Although many variations and combinations can be used, there are only three basic types of layout:

1 A singular drainage system is described often as a **fen layout**. Each drain has direct outlet to an open channel as shown on the left of Fig. 15.3.
2 A composite system with laterals meeting the leader at right angles is called a **grid layout**. Lateral drains may connect to the leader on one side only or on both sides depending on local topography and the position of the leader drain. A grid layout is illustrated on the right of Fig. 15.3.

Figure 15.3 *Layouts for flat land*

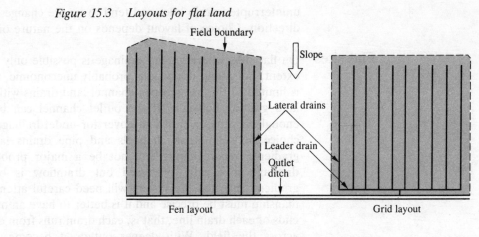

Figure 15.4 *Herringbone layouts*

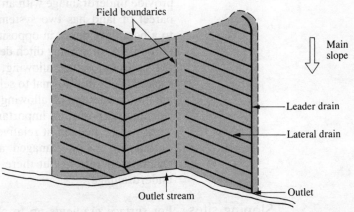

3 **A herringbone layout** differs from a grid layout only in that the drains join at angles of less than 90°. The angle selected depends on the direction needed to obtain the optimum gradient for lateral drains. Typical layouts are shown in Fig. 15.4.

Choice of layouts Layouts may be used singly or in combination depending on the complexity of landforms. It is essential to be flexible for the more complex sites. Precisely laid parallel drains can have correct alignment at one end of a site but not at the other end if direction of slope changes across the site. Drainage contractors prefer long uninterrupted drain lines and tend to ignore changes of gradient direction. Choice of layout depends on the nature of the site.

Flat sites

On flat sites where gravity drainage is possible only to a limited extent, but pump drainage is probably uneconomic, the designer is limited to the use of open-channel land drains with or without land grading between. If the outlet channel can be just deep enough to provide sufficient cover for underdrainage, there is a choice between open channels and pipe drains laid with flat gradient. Where silting will not be a major problem, underdrainage is usually preferred but drainflow is by hydraulic gradient only and the system will need careful attention. Workmanship must be precise and it is better to have an outlet at both ends of each drain line, that is, each drain runs from ditch to ditch across the field. With deeper outlets it becomes possible to provide underdrainage with an actual gradient. In most cases each parcel of land has two systems of underdrainage, each running to a boundary ditch, in opposite directions, from the midway line in the field. Increasing ditch depth permits greater gradients up to the optimum value, allowing, in turn, larger field sizes. On all such flat sites it is normal to select a singular drainage layout. This has the advantage of allowing easy access to each drain line for cleaning purposes, an important consideration since maintenance work will be needed at relatively short intervals. The many drain outlets are easily damaged and can be a nuisance for ditch-cleaning operations but there is little alternative for trouble-free underdrainage.

Sloping sites

For surface gradients up to about 0.5 per cent – depending on operator skill – the drains must be aligned exactly downslope to achieve the best drainflow characteristics. For greater gradients it becomes necessary to angle drain lines across the slope to obtain a suitable gradient and the best possible interception of any downslope water movements either on the surface or through the profile. On steeper slopes drain lines should cross the slope with gradients of about 1–2 per cent which allows some margin of error for less than precise gradient control. Drains aligned with a major slope can have high drainflow velocity and fail to drain the land between drain lines unless drain intervals are very small. For sloping sites the designer can choose between singular and composite layouts. Unless there are special problems such as ochre formation or rapid siltation which would require frequent and easy access to the pipes, it is better to have the minimum number of outlet points. The few major outlets can then be substantial constructions not easily lost or damaged. This implies selection of a composite layout wherever possible. On the least sloping sites the grid layout allows easy cover of the whole site but undulations and steeper gradients usually require selection of a herringbone layout.

Drain spacings

Selection of the interval between drain lines in a regular layout is perhaps the most discussed topic of drainage design. Generally speaking, drains that are too far apart will fail to drain the soil profile adequately mid-way between drains while those that are set too closely, although successful, are unnecessarily expensive. Even so, it would be a mistake to attach too much importance to this facet of drainage design. Unless choice of spacing is grossly wrong the results are usually only marginally below optimum and drain spacing is blamed for many drain failures which are actually the result of other design faults, such as omission of permeable fill, or poor workmanship. It is more important to diagnose correctly the drainage problem or problems and to apply the most effective drainage solution. When this is done correctly the choice of drain spacing is not a precise calculation. All that is necessary is that spacings are of the correct order and, as a general rule, selection can be confined to one of a few standard intervals from the range 10 m, 15 m, 20 m, 25 m, 30 m, 40 m, 50 m and 60 m.

Drain depth

The depth chosen for underdrainage is limited by site topography and the minimum permissible soil cover to prevent damage by field traffic. Where there is an appreciable range of depths available the depth chosen must relate to the drainage problem. Since depth and spacing are not entirely independent factors it is most convenient to discuss them together in terms of layout design.

Classifying drainage problems

For most practical purposes new drainage systems are required to solve one or more of the following basic land-drainage problems:

- A high groundwater table in a surface aquifer
- A perched water table in a surface aquifer
- A perched water table in a surface aquiclude
- Overland surface flow
- Groundflow and spring seepage
- Miscellaneous problems presenting special difficulties.

These are discussed below in terms of drain depth, drain spacing and any other special design features that may be required. For a designer in the field, in addition to positive identification of the primary and any secondary drainage problem, it is necessary to establish whether the profile is effectively a surface aquifer or a surface aquiclude. This is a matter of classifying the soil horizons in terms of texture and structure.

Classifying profile permeability

For choosing practical drainage solutions, soils may be grouped into three broad permeability categories by means of a textural classification.

1 Freely permeable surface aquifers occur where the profile to drain depth is composed of the following texture classes: coarse sand; loamy coarse sand; coarse sandy loam; sand; loamy sand; sandy loam.
2 Moderately permeable surface aquifers occur where the profile to drain depth is composed of the following texture classes: fine sand; loamy fine sand; fine sandy loam; very fine sand; loamy very fine sand; very fine sandy loam; loam; silty loam; silt.
3 Slowly permeable and impermeable surface aquicludes are found where the profile is composed of poorly structured or massive horizons of the following texture classes: sandy clay loam; silty clay loam; clay loam; sandy clay; silty clay; clay.

Wholly indurated profiles of any texture behave as aquicludes. Where an individual horizon is compacted or indurated this also has the effect of making the profile an aquiclude. But such compaction of the profile should be disrupted as part of a drainage project and the influence of texture will be restored. Fissuring in a profile caused by soil structure can increase permeability, sometimes markedly so, but it does not convert an aquiclude into an aquifer in terms of flow lines towards drainage channels unless the fissures are permanent and penetrate the profile to drainage channel depth. The basic drainage problems can now be discussed in turn.

Controlling a high groundwater table in a surface aquifer
Design requirements

Soils can withstand field traffic without damage provided that the zone of saturation is at least 50 cm below the surface. For surface aquifers this means the position of the groundwater table. Drainage design requirements are to control the rise of the groundwater table during periods of rainfall so that, at mid-drain points and at design drainage flow rates, the groundwater table will not be less than 50 cm from the surface for any period exceeding one or perhaps two days, certainly less than 5 days. Referring to Fig. 9.2, the maximum groundwater table height between drains is determined mainly by rate of recharge (R), profile hydraulic conductivity (K), drain depth ($H + 50$ cm) and drain spacing (L). To a lesser extent it is also influenced by concentration loss and by the depth to the impermeable base of the aquifer (D) provided this is near enough to impede flow lines. Recharge rate is the chosen design drainage rate, concentration loss depends on the type of pipe used and in most cases can be ignored while the value of D can be observed in a test hole. It remains then to obtain a value or values for K in order to select the most suitable drain spacings.

Assessing profile hydraulic conductivity

The most practical and convenient method of assessing hydraulic conductivity values is by careful examination of all soil horizons above and below drain depth as described in Chapter 4, referring to Fig. 4.6.

Measuring profile hydraulic conductivity

Several methods of direct measurement have been described but results are reliable only in uniform profiles that possess textural macropores and have a high groundwater table. In other situations measurement can be very inaccurate and, where used, the calculated results must be reasonably compatible with observed profile characteristics. A satisfactory single auger hole method is described by Trafford in the UK Ministry of Agriculture Technical Bulletin 74/7.

A truly vertical auger hole is opened deep enough to pass below the groundwater table and left for a day to allow return of stable conditions. The depth of water in the hole is then measured. The water is baled out and careful note is taken of the rate of rise of water in the hole against a fixed reference point using a float on the end of a measuring tape. Using only the lowest quarter of the depth of the hole the rise of water level is noted against standard time intervals of 5, 10 or more seconds depending on the rate of inflow and the process is repeated until steady values are obtained. For calculation it is also necessary to know the depth to any relevant impermeable base of the aquifer.

If K is hydraulic conductivity in m/day.

H is the depth of hole below the groundwater table in cm.

Y is the average distance between groundwater table and level of water in the hole in cm.

Δt is the selected time interval in seconds.

ΔY is average change of water level in standard time in cm.

r is the radius of the hole in cm.

Z is the depth in cm to an impermeable layer below the bottom of hole.

Then, on sites where $Z > \frac{1}{2} H$, hydraulic conductivity can be calculated as:

$$K = \frac{4\,000\,r^2}{(H + 20r)\,(2 - Y/H)\,Y} \frac{\Delta Y}{\Delta t}$$

If the auger hole reaches an impermeable layer and $Z = 0$ the formula becomes:

$$K = \frac{3\,600\,r^2}{(H + 10r)\,(2 - Y/H)\,Y} \frac{\Delta Y}{\Delta t}$$

For intermediate conditions where $Z < \frac{1}{2} H$ but > 0 there is no suitable formula. Use both formulae for calculating $\acute{K}$ and choose an intermediate value.

Choosing drain depth and spacing

Many mathematical solutions have been proposed for calculating drain spacings. Most are extremely complex and of use only where values of the various factors can be obtained in practice. The simplest and most widely used is that of Hooghoudt. Referring to the notation used in Fig. 9.4. the Hoodghoudt formula may be stated as:

$$L^2 = \frac{8K_2 dH}{R} + \frac{(4K_1 H^2)}{R}$$

where L is drain interval in metres
K_1 is profile K above drain level in m/day
K_2 is profile K below drain level in m/day
H is groundwater table height above drains in m
R is the chosen design drainage rate in m/day
d is a function of D in m.

This formula assumes that profile hydraulic conductivity above drain level is uniform but differs from profile hydraulic conductivity below drain level, which is also uniform. This, of course, is a fairly artificial situation but in practice is not too serious. In layered profiles where the junction between the two zones of assumed uniform hydraulic conductivity does not more or less coincide with the level of drains it is better to assume a single K value ($K_1 = K_2$) for the whole profile using the lower K value.

Depth to an impermeable layer below drain level (D) has a direct effect on the solution for L up to a maximum value of about 10 m for D. It may be used directly for flow towards ditches but for underdrainage any flow lines below drain level are more restricted and it is better to use a smaller value (d), the depth equivalent as shown in Fig. 15.5.

Since d is a function of both D and L the formula cannot be used directly for computation of L. It is necessary to assume a value for L to obtain d, then calculate L and repeat until there

Figure 15.5   Equivalent depth values

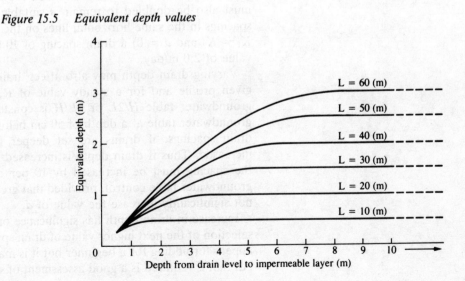

is a reasonable match between assumed L and calculated L. In a homogeneous profile where $K_1 = K_2$ the formula can be simplified:

$$L^2 = \frac{K}{R} [8dH + 4H^2]$$

All such methods are very tedious for a practical drainage designer and it is preferable to prepare a ready reckoner in advance to cover most situations. One such system is described by Carter and Trafford in UK Ministry of Agriculture Technical Bulletin 73/12. By choosing fixed values for H (0.5 m) and R (0.01 m/day) and by selecting only ten ideal profiles with various combinations of values for d, K_1 and K_2 it is possible to show the relationship between such values and the selected drain spacing L as in Table 15.1. Where field conditions approximate to the selected values of H and R, the table can be used directly. Values of D and K in the profile are observed, d is assessed and these values applied to the table. Select the best match, by interpolation if necessary, in the vertical columns to obtain the best drain spacing.

For different values of R the relationship between L and K will be different. But, since $K/R = $ a constant, the table can still be used by notionally adjusting the K values. For example, if R is doubled from 0.01 to 0.02 m/day then the K values on the table must also be doubled to remain a suitable match for the drain spacings in the same horizontal lines on the table. That is (where $K_1 = K_2$ and $d = 0$) a drain spacing of 10 m would require a K value of 2.0 m/day.

Varying drain depth may also affect drain spacings since, in a given profile and for a steady value of R, the gradient of the groundwater table $H/2L$ or $2L/H$ is constant. To maintain the groundwater table at a depth of 50 cm below the surface at mid-drain spacings, if drains are set deeper the spacing must be increased. Thus if drain depth is increased by, say, 20 per cent, the spacing must be increased by 10 per cent to get the same groundwater table control, provided that greater drain depth does not significantly increase the value of d.

Increase in drain depth has significance only where it warrants selection of the next higher value of drain spacing. All of this may appear forbidding for a beginner but it is manageable in practice. All that is required is a good assessment of soil texture and struc-

Table 15.1   Selecting drain spacings for K values in m/day

L (m)	$K_1 = K_2$					$K_1 = 2K_2$			$K_1 = 5K_2$		
	$d = 0$ (m)	$d = 0.5$ (m)	$d = 1.0$ (m)	$d = 2.0$ (m)		$d = 0.5$ (m)	$d = 1.0$ (m)	$d = 2.0$ (m)	$d = 0.5$ (m)	$d = 1.0$ (m)	$d = 2.0$ (m)
10	1.0	0.4	—	—	Upper Layer (K_1)	0.5	—	—	0.7	—	—
					Lower Layer (K_2)	0.25			0.1		
15	2.3	0.8	—	—	K_1	1.1	—	—	1.6	—	—
					K_2	0.6			0.3		
20	4.0	1.3	0.9	—	K_1	2.0	1.5	—	2.9	2.5	—
					K_2	1.0	0.8		0.6	0.5	
25	6.3	2.1	1.4	—	K_1	3.1	2.3	—	4.5	3.8	—
					K_2	1.6	1.2		0.9	0.8	
30	9.0	3.0	2.0	—	K_1	4.5	3.3	—	6.5	5.3	—
					K_2	2.3	1.6		1.3	1.1	
40	16.0	5.3	3.5	2.0	K_1	8.0	5.7	2.5	11.5	9.3	5.0
					K_2	4.0	2.9	1.3	2.3	1.9	1.0
50	25.0	8.3	5.4	3.0	K_1	12.6	9.0	5.4	17.9	14.5	10.2
					K_2	6.3	4.5	2.7	3.6	2.9	2.0
60	36.0	12.0	7.8	4.4	K_1	16.0	12.8	7.8	25.7	20.9	14.8
					K_2	8.0	6.4	3.9	5.1	4.2	3.0

Notes: 1. A value of $d = 3.0$ relates only to 60 m drain spacings and can be ignored.
2. If $K_1 > 5K_2$, ignore K_2 and assume $d = 0$.
(Based on MAAF Technical Bulletin 73/12)

ture, observation of aquifer depth and the application of these values to the table making allowance for different design drainage rates. Spacing needs to be increased less often for any significant increase of drain depth.

Practical application of methods for selecting drain depth and spacing

The methods described above apply only to surface aquifers affected by a high groundwater table and are suitable for the whole range from the least permeable materials requiring drains at 10 m spacings to the very permeable profiles that can be drained at intervals of 60 m. In all cases drains should be laid at greatest possible depth as determined by outfall level. Usually the depth is restricted by outfall limitations where there is a high groundwater table with most drains being laid within the range of 60–90 cm of drain cover. The most likely exceptions would be terrace sites or other plateau areas with an aquifer trapped behind a barrier to lateral groundflow. In such cases drains should not be set impossibly deep but any practical level for breaching the retaining barrier can be selected allowing drains to be as much as 1.5 m below the surface. Setting drains at greater depth brings the risk of over-draining profiles which may already have a deficiency of available water for crops.

Once drain depth is determined the drain spacings can be selected as described above. This method of choosing drain depth and spacing assumes that there will be no associated soil treatment or use of permeable fill. In the least permeable aquifers there is a chance that the drainage problem will be a perched watertable, particularly in a wet climate or where the profile has slumped – such profiles contain a high proportion of unstable particle sizes. In such cases soil treatment and permeable fill may be necessary, in which event the methods described above do not apply and the solutions described for a perched watertable must be used instead. In any situation where tests or observations indicate that drain spacings of less than 10 m are needed the solutions for draining an aquiclude must be used.

Controlling a perched water table in a surface aquifer

There are two causes of a perched water table in an aquifer. Downwards groundflow may be inhibited by natural induration – for example, by the presence of a pan – in any type of profile or, as mentioned above, rainfall intensity may exceed percolation rate where naturally unstable particles in a profile become slumped and compacted. Where a pan is identified it must be disrupted by subsoiling. Underdrainage may not be needed if the groundwater table does not come near the surface during the critical season. Where underdrainage is needed its design should be based on the requirements of a high groundwater table as described above for unimpeded aquifers. Subsoiling will have removed the potential to form a perched watertable. Where percolation is inhibited by an unstable medium texture that has become compacted it is necessary to subsoil regularly over drains connected to the topsoil by permeable fill. A nominal drain spacing of 20 m is satisfactory so long as soil management is of a high standard. Where drains are set closer than 20 m (i.e. at 10 or 15 m) it may be sufficient to place permeable fill in every second drain trench. Drains should be at minimum permissible depth to economise on permeable fill.

Controlling a perched water table in a surface aquiclude

Surface aquicludes may result from the presence of fine-textured profiles or because of massive induration of other texture classes. Drainage techniques for improving fine-textured, slowly permeable soils are described in Chapter 12. Drain spacings for underdrainage is a function of the persistence of fissures or channels formed in the profile by soil treatment. Suitable spacings range from 20 m upwards in steps of 10 m. It should be remembered that the piped drainage channels are acting more as collector drains than as field drains. Permeable fill is always needed to connect the zone of profile fissuring or the unlined channels to the pipe drains which must always be set at a minimum depth. Massively indurated profiles that are freely permeable when disrupted need successive subsoil treatments over an underdrainage system at nominal 40 m spacings without permeable fill. Since subsoiling never reaches drain depth the profile is not converted into a surface aquifer in terms of flow lines towards drains. For this reason the drain spacing table for surface aquifers does not apply. For intermediate soils – indurated profiles that are potentially only moderately permeable – it becomes necessary to consider closer spacings and permeable fill as described for perched watertables in aquifers.

Controlling overland surface flow

Surface flows are most conveniently controlled by open channels but where underdrainage is used it must have connection right to ground surface by means of the larger grades of permeable fill or pebbles. Piped drains can be used to control small surface flows such as those arising from a spring line situated in a steep bank that cannot be drained directly. The usually single drains must be set along the upper margin of the land to be protected and at minimum permissible depth.

Controlling spring seepage

The fundamental requirement in controlling lateral or upwards groundflow movements is to intercept the flow before it can saturate the plant rooting zone in a soil profile. In nearly all cases the work will require the careful siting of interceptor drains.

Gravity spring seepage

The majority of gravity seepages take the form of a spring line across a slope. The best site for an interceptor drain is not at the spring seepage site but someway upslope from it depending on the slope of the surface and the nature of the rock structures causing the spring. Figure 15.6 indicates the best position for a interceptor drain in an easy pathway spring. The drain must be set on the impermeable base of the aquifer and must be protected by a minimum depth of soil. To meet both requirements the drain must be placed sufficiently far behind the spring to obtain full cover in the aquifer. Where the base of the aquifer and ground surface meet at a very small angle it may be necessary to sink the drain some way into the basal aquiclude and it is then worthwhile to use permeable backfill up to the level of the aquifer. The best siting for drains to control any overtopped barrier springs is in the aquifer immediately behind the barrier as shown in Fig. 15.7. In

Figure 15.6   Intercepting ground flow

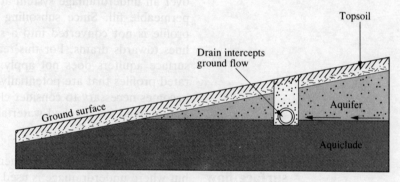

Figure 15.7   Intercepting seepage behind a barrier

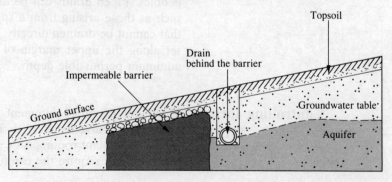

all cases the drain must be on the upslope side of the barrier. A series of individual outcropping aquifers or closely set barriers will need a separate interceptor drain for each spring line.

Artesian spring seepage

Artesian springs present different problems and the method adopted must be specifically matched to the rock structures present. Diffuse artesian seepage into or through a semi-confining layer may resemble a patchy, topographical high groundwater table but the groundwater behaves differently and drains set in the semi-confining layer will not be effective because of the flow-line pattern which is as shown in Fig. 9.6. The upward flowing water bypasses drains to reach the surface. Where the aquifer is below a totally confining layer, groundwater will reach the surface at only a few well-defined points and is more easily identified. A situation not easily recognised on the surface is a confined aquifer structure with a piezometric surface at or just below ground surface so that no obvious upwelling occurs. All types are easily recognised in deep test holes. In all artesian springs, drainage of affected soil horizons depends on controlled release of confined groundwater through interceptor drains. There are three distinct situations:

1 The confined aquifer can be reached at the spring site by drains graded from the available drain outlet point.
2 The aquifer is below the depth of a possible drain at the spring site but can be reached by an excavator digger.
3 The aquifer is deeper than the reach of an excavator digger at the spring site.

Setting drains within a confined aquifer

The most simple situation, and fortunately the most common, occurs where interceptor drains can be laid in the confined aquifer. If the aquifer is highly permeable and uninterrupted over the whole site, a single, large capacity drain (perhaps 15 cm diameter) may be sufficient to release the confined groundwater pressure. However, the aquifer is usually of more limited permeability or is interrupted by ribs of undecayed or non-jointed rock and a number of interceptor drains will be needed for comprehensive improvement. Because the higher springs on a hillside may be gravity seepages, it is normally necessary, to lay a pattern of drains across the slope with the highest drain passing along the upper margin of the wet area. Where hard rock projections do not disrupt the pattern the drains can be laid at regular spacings of 30–80 m depending on the slope and permeability to the aquifer. Spacing is not critical and where it is evident that the pressure has not been fully released, an extra drain can be laid. Release of pressure in the aquifer will stop upwelling at the springs even where drains do not pass close to the sites of rises, but a few widely separated rises may be drained directly with individual drains leading to the site of upwelling. The best location for drains in the profile is shown on the left side of Fig. 15.8.

Figure 15.8 Draining a confined aquifer

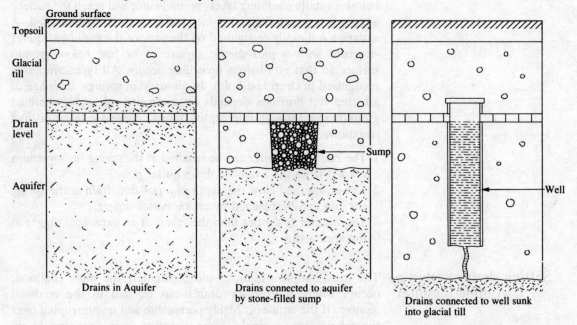

Ground surface

Topsoil

Glacial till

Drain level

Aquifer

Sump

Well

Drains in Aquifer

Drains connected to aquifer by stone-filled sump

Drains connected to well sunk into glacial till

Draining deeper aquifers

Where drains graded from an outlet cannot reach a confined aquifer, because the site is too flat or because the confining layer is too deep, they must be set in the confining layer and will be effective only if connected to the aquifer by permeable sumps. Drains should be laid as deep as possible. A hole should be excavated down to the aquifer within the improvement area at intervals of about 40 m along each drain trench, then filled with stones or gravel as illustrated in the centre of Fig. 15.8. This is sufficient to release the confined groundwater. Very often, in larger projects only drains at the lower end of the site need to be sumped. The drain layout is as described on p 220 for drains set in an aquifer.

Sites with very deep confining layers

Where confining layers cannot be easily breached by the larger type of excavator digger, artesian springs are not common. Where they do occur land improvement is difficult. The very high pressure needed to penetrate the confining layer creates very active springs which cause considerable upwelling and groundwater spreads over the surface converting the whole area into a mire. Successful drainage requires identification of each rise. This cannot be done until surface water is cleared. It may be achieved by setting out a regular system of drains, usually in herringbone pattern, over the site and spaced at about 40 m with the first drain laid along the upper margin of the wet area. Other drains are laid in turn downslope to clear surface water as work progresses. The site should then be left to dewater into still open drain trenches after which spring rises are more easily identified. Each spring rise should then be excavated to the full reach of the digger. Often an ochre-stained vertical channel can be followed downwards through the confining layer. A well should be constructed over the spring rise using pipe sections as available but 30 cm diameter usually is adequate. This should then be connected to the system already laid by a short length of drain and the whole construction can be as shown on the right of Fig. 15.8. The well may be capped and left below the soil surface if desired. Once all springs are connected to the interceptor drains and the site has dried the whole system may be backfilled.

As discussed in Chapter 11, there is no practical method of easily assessing flow rates without lengthy periods of flow measurement. In most cases artesian drainage is satisfactory if pipes of adequate diameter are selected arbitrarily. A minimum size of 100 mm diameter should be used for interceptor drains with 150 mm or 225 mm leader drains.

Special difficulties associated with spring issues

Points requiring special care with spring issues are as follows:

- Moving groundwater must always be given direct connection to the drains. This is particularly true for artesian seepage when drains must be set in the confining layer. Water seeping upwards or upwelling through a vertical channel to the surface cannot easily flow laterally through the profile towards a drain even if it is less than a metre away.

- Artesian seepage passing through a confining layer with a high clay content will usually have produced a zone of slurried clay which moves slowly upwards as a viscous liquid. Any drain laid through the unstable area is sealed by clay and pushed upwards out of alignment. It is necessary to excavate all the slurried material.

- Clay-rich confining layers are often slurried along the contact surface with the confined aquifer. The slurried clay can cause difficulty when the drain trench is excavated. Release of pressure may allow slurried clay to flow into and upwards in the open trench upheaving the drain pipes. Only slightly less disconcerting is the tendency for slices of confining layer material to slide sideways on the slurry to block the trench and disrupt the drain alignment. Where this happens the trenches need to be left open to allow a period of dewatering.

- Highly permeable aquifers of loose rock waste can be so charged with water that they behave like a fluidised layer and flow down the open drain trench from the point where the aquifer is first exposed. This sweeps all the pipes before it and can be dangerous for workers in the trench. The outflow leaves a cavity below the confining layer which then collapses in large sections, leaving a large hole. A period of dewatering allows release of pressure and when work can begin again it is usually found that the drainage problem has been solved.

- All deep excavations are dangerous and care is needed in a deep drain to ensure that side walls do not collapse on workers.

- Drainage work to control springs must be based on a flexible approach that takes account of changing circumstances as work progresses. Very often the trench excavation reveals new features which must be taken into account. For this reason the work can only be done by an excavator digger forming a wide, open trench. Continuous trenching machines and trenchless drainers are quite unsuited to the work as are drainage contractors with fixed methods of operation.

Miscellaneous problems presenting special difficulty

Drainage work in river flood plains

River flood plains are complex and delicately balanced structures which require careful consideration before any drainage improvements are attempted. The dominant land drainage problem is a high groundwater table; but there are other complications. The following are the ones frequently encountered:

- Flood plains are more or less level in a transverse direction so that drains cannot be laid to run water directly to the river. If the valley has sufficient longitudinal (down-valley) gradient or if the valley is stepped, with a steeper section not too far away, the necessary gradient and cover for drains may be obtained by selecting an outlet point far enough downstream. On a small scale the flood plain drainage may be improved by blasting a deeper channel through the local base level-forming rocky outcrop. Where gravity drainage is not possible the feasibility of pump drainage must be examined.
- All flood plains can be flooded in times of swollen rivers. Arable crops are always at risk unless protected by flood banks.
- Before any work is done it is necessary to ensure that the groundwater table can be controlled. This will not be possible if groundwater in the flood plain has free connection with water in the main channel. Observation of water levels in test holes is necessary over a period of time to see if there is a response to changes of water level in the channel. Where return groundflow is rapid very little can be done to improve the site other than to lower the level of water in the channel, which in itself is rarely possible. Slower return rates may be overcome by pump drainage but operating costs are high.
- Artesian springs may rise from depth anywhere in the flood plain. They require special attention. Gravity seepages, surface run-off or drain outlets may cause further inflows along the base of marginal bluffs. Interceptor channels, either drains or ditches, will be needed to control these additional inputs of water.
- The nature of the soil profile is important. Alluvium usually forms a moderately permeable aquifer which tends to be very unstable. It also may be a semi-confining layer for the underlying gravel and the piezometric surface may rise and fall with the level of water in the channel. Thus from place to place and at different seasons the drainage problem may range from high

groundwater table to perched water table or semi-confined artesian seepage. Drain depth will be severely restricted so that, in most cases, drains will be set in the alluvium. It is usually found that 10 m or 15 m spacings will be required and the need for permeable fill above drains and sumps below the drains will be a matter of judgement.

- The condition of the main channel is important. Where it is stable with well-formed banks covered by established vegetation, or where engineering works have secured the banks, drain outlets should be safe. Where lateral erosion is clearly active there is always the danger that drain outlets will be swept away or buried by deposited rock waste.

- Because of inherent difficulties, high installation costs and the ever present danger of flood damage, it is necessary to consider very carefully the cost/benefit aspects of flood plain drainage.

Drainage work in organic soils

Drainage of wholly organic soil profiles presents special problems that need to be considered at design/project feasibility stage. These are:

- The unstable nature of saturated peat prevents the use of machinery on the surface. Peat can be stabilised sufficiently by means of preliminary dewatering which involves opening up ditches as and when conditions allow. Even where direct installation of underdrainage is possible the work may fail for other reasons associated with the presence of slurried peat.

- Choice of drain-laying machine is determined by the nature of the soil profile. Even in good climatic conditions and following satisfactory dewatering, it is necessary to use only track-laying equipment with a low ground pressure. The presence of buried logs and stumps may prohibit use of specialised drainage machines or at least require the presence of an excavator/digger.

- Drained peat always shrinks and the lowered surface may cause loss of gradient. Allowance for shrinkage must always be made at the design stage and newly installed drains must be at least 1 m deep, even if this renders pump drainage essential.

- Peat shrinkage can affect the profile below drain level and the degree of shrinkage is by no means uniform. Pipe alignments and gradients can be distorted and action must be taken to reduce the effects. Continuous lengths of plastic piping are more satisfactory than clay tiles. Use of permeable fill above the pipe is valuable for allowing continued flow across a sag in the pipe.

- Organic profiles are 'spongy' and retain water more tenaciously than many mineral soils. Furthermore, the hydraulic conductivity of the profile tends to be reduced after drainage improvement and this must also be considered. Where arable cropping is intended, drain spacings must range from 10 m intervals for massive profiles up to no more than 20 m for drummy peat profiles.
- Saturated peat is highly mobile and when disturbed by pipe-laying operations it can flow into pipe slots and seal off the drains. Care is needed to ensure that the profile at drain depth is stable and the use of permeable backfill is of considerable benefit.
- Even in dry conditions pipe laying is not without difficulty. Fragments of peat float in the groundwater and flow towards the drains where they are trapped in pipe slots and block them before the trench can be filled to act as a filter. Trenchless drainage avoids this problem provided the surface can support the machine.
- When all of these problems have been solved the reclamation may still fail because of the accumulation of iron ochre in and around the pipes.
- Because of the many practical difficulties the initial cost/benefit assessments must be thorough.

Drainage works to cope with iron ochre

As discussed in Chapter 4, certain soils can cause trouble when drained because profile aeration allows the precipitation of iron ochre in pipes and ditches. It may take the form of floating gelatinous globules or it may be deposited as hard scale in and around drain pipes or in the soil nearby. In the worst cases and where financial returns are likely to be limited because of climatic factors, expenditure on underdrainage may not be worthwhile. If drainage is considered worthwhile, the following precautions can help reduce the risks:

- Assessment of the most difficult areas within a site should be made before work begins. This may be identified as a peat hollow or change of soil type; where this occurs the most difficult area can be abandoned or drained with separate outlet.

- Wherever possible, use singular layouts to provide easy access to each drain. Where topography makes this impossible the layout should have as many junction and inspection chambers as possible.

- Gelatinous ochre is easily flushed out of pipes by jetting. If it is left inside a pipe during a dry season it is converted to an immobile scale which cannot be removed. For this reason each drain should be jetted thoroughly each year at the onset of the dry season to prevent scale formation. Where scale has formed a high pressure type of jetting equipment must be used.

- Regular and deep subsoiling in dry conditions will encourage the oxidation of iron compounds within the profile where no harm is done. For the same reason a preliminary dewatering with open channels may remove some of the ochre before drains are laid.

- Certain techniques have been employed to inhibit ochre-forming bacteria. Most are either impractical or ineffective but a very promising method has been described by the Macaulay Institute for Soil Research in Scotland and involves causing the drainage water to pass through filter units packed with pine bark.

- If hard scale forms directly inside the pipe, around entry slots and in soil or permeable fill nearby there are no practical measures available to reactivate the sealed off underdrainage system. If the underdrainage is replaced it is always at risk of further ochre formation which therefore will be economical only on the most fertile soils.

- When installing new schemes care should be taken to set tiles with a gap of about 2.0 mm or to select plastic piping especially designed for the purpose with 2.0 mm wide slots. This greatly reduces the incidence of slot blockage. Filter wrap pipes – those inside a sleeve filter – must never be used.

Drainage works in unstable soils

The basic difficulty in draining profiles composed of the least stable particles is that the particles are easily removed from the profile, especially from the disturbed soil above the drain which then move into the drain with the drainage water. Once inside the pipe the particles form a smooth homogeneous layer which is difficult to erode and the pipe becomes blocked. The problem resembles the effects of ochre formation in a number of ways and in the most difficult cases underdrainage may not be worthwhile. There are a number of factors that must be considered if underdrainage is to be successful:

- Easy access is needed for drain cleaning and singular layouts are to be preferred. If this is not practicable, as many junction and inspection chambers as possible must be installed and silt traps regularly inspected.
- The smallest possible entry slots must be used. Clay tiles must be set as closely as possible and plastic pipes must have the narrow 1.0 mm slots. If wide-slotted pipes are employed even the larger sized sand grains can enter the pipe.
- The pipes must be cleaned out regularly by rodding or jetting, but where jetting is used it must be carried out with care.Only low pressure systems are satisfactory since high pressure jetting can cause soil scouring outside the pipe.
- Filtering devices may be considered. All types of filter, to be most effective, must have a mesh size that just matches the size of particle to be removed. When this is achieved the filtering effect is excellent until all meshes are blocked and the filter becomes so impermeable that it seals off the pipe. At this stage it would be ideal if the filter could be changed but this is rarely possible. Less efficient filters remain permeable for a longer period and may provide some protection until the soil above the pipe becomes more stable. Filtering devices take several forms:
- The traditional method was to set aside the turf from the drain track and to invert it on top of the pipe. Similar systems involve the use of straw or peat litter to act as a filter. Most have some effect until the organic fibres decay but this may be sufficient to establish the underdrainage system.
- Fine gravel may be used but it is not a very good filter and is too expensive for use simply as a filter.
- Proprietary filter-wrapped pipes of various types are available and these work well at first. In the best conditions they work until the soil stabilises and all is well – in the worst they become blocked by particles and the system must be replaced.
- A compromise solution is the use of filter mats on top of the pipes. These operate just like the fully wrapped pipe but in the

most difficult conditions it may be possible to replace them without disturbing the pipe.

- The use of filters is an uncertain technique and no general guidelines are available for their satisfactory use. One point which is very clear is that even small amounts of ochre render them impermeable. Thus fine sands and silts which also have an inherent ochre problem are particularly difficult soils to drain.

Part three A systematic approach to field drainage

Steps in designing a drainage scheme

Drainage problems are more easily solved at the least possible cost if the work is tackled in a logical sequence. With so many factors to consider it is easy to forget a step and to make a design decision before all of the relevant information is available. Anyone with a fixed method of drainage for all types of site has fewer design problems but then the work is often unsatisfactory or unnecessarily expensive. Every facet of drainage design must be considered for every site and it is necessary to develop an organised, systematic approach which is comprehensive. It will be found that a check list is very helpful. For any site the survey, design and supervisory work can be tackled in a sequence similar to the one suggested below.

Preliminary Survey Examine the site for evidence of drainage limitations and seek any evidence of crop performance over recent years. [ref Ch. 2].

Assess the likely benefits to be obtained from improved land drainage. This requires estimating additional sales resulting from the improvement and deducting additional trading purchases and production costs to arrive at likely gross margins as described by agricultural economists. Depending on the results:

Make or seek a decision to proceed with a detailed survey
This will involve the cost of using or hiring an excavator/digger and may also require a detailed levels survey.

Detailed survey Prepare a sketch plan of the site and surrounding area showing field boundaries, roads, power lines, known underground pipelines and any other features of significance for drainage work.

Examine all nearby ditches and streams and note their condition. Mark these open channels on the sketch plan showing direction of flow. [ref. Ch. 8 and 10]

Measure the surface gradients on the site and mark them on the plan showing the direction of slope and the best outlet point. For very level sites, especially where gravity flow drainage is not certain, it may be necessary to arrange for a detailed levels survey and this may include the preparation of a contour map with contours at 30 cm intervals.

Identify and sketch in all significant landforms within and surrounding the site. [Ref. Ch. 6 and 8]

Note the presence of surface inflows from outwith the site. [ref. Ch. 6 and 15]

Identify the general distribution of topsoil types within the area using an auger or spade. Indicate the boundaries of different soil types on the sketch plan. [Ref. Ch. 4 and 5]

Attempt to identify the nature of underlying rock formations and drift taking account of soil type, landforms and any nearby rock outcrops. Consult geological map if available. [ref. Ch. 3, 6 and 9]

Open up a sufficient number of test holes across the site to reveal profiles of the soil types present and any underdrainage system. A number of deep test holes are needed also to examine deeper layers.

Note the nature and condition of existing underdrainage systems. [ref. Ch. 11]

Note the relevant soil properties including evidence for poor drainage. [Ref. Ch. 4 and 5]

In the deep test holes note the depth to any less permeable barrier below the soil horizons. [ref. Ch. 9 and 15]

Note the level of any subsurface aquifer and whether groundwater issues from it when exposed. Note the level reached by the issuing groundwater in the test hole. [ref. Ch. 9 and 15]

In all test holes note any gradient in the groundwater table between individual holes and in nearby waterways (in surface aquifers only). [ref. Ch. 9 and 15]

Carefully assess the hydraulic conductivity of the soil horizons both above and below likely drain levels in the profile. [ref. Ch. 4, 5 and 15]

From all of the evidence identify the main drainage problem and any secondary drainage problem. [ref. Ch. 1 and 15]

Maintenance work

Where the evidence suggests that a marked improvement will be achieved simply by improving the existing system this should be carried out before any other drainage work is attempted.

Make or seek a decision to invest in maintenance works
Supervise the work as it proceeds. Examine improved ditch banks for field drains and improve as required; pit and rod or jet out under-drainage systems as required; subsoil to disrupt pans as required. [ref. Ch. 10, 11 and 12]

Drainage design

Where it is clear that new drainage works are needed, proceed with drainage design.

Decide which outlet channels need preliminary improvement or where new outlet channels are needed. [ref. Ch. 8 and 10]

Select the best drainage solution for the identified drainage problem(s). [ref. Ch. 1 and 15]

Select the most suitable design rainfall rate. [ref. Ch. 7]

Calculate the peak discharge rate for ditches. [ref. Ch. 10]

Calculate the design drainage rate for underdrainage. [ref. Ch. 11]

Assess or calculate drainflow rates for interceptor drains. [ref. Ch. 11 and 15]

Select the best layout for outlets (pump or gravity), open channels and underdrainage works. [ref. Ch. 12, 13, 14 and 15]

Calculate pump capacity, ditch dimensions, pipe sizes as appropriate. [ref. Ch. 10, 11 and 14]

Calculate the amount of materials required and estimate the total costs.

Assess the worthwhileness of the project. Compare the gross margins achieved from the improvement with the capital costs of the improvement. In most cases the capital sum should be recovered in a period not exceeding 20 years. Where pumping costs have to be met they must be deducted from the gross margins.

Make or seek a decision to drain comprehensively, to abandon or to attempt a low cost improvement

If drainage is to be carried out: Prepare a detailed plan and specification of the proposed works. Consult environmental or planning authorities as appropriate, arrange wayleaves as required, make adjustments to plan and specification to comply with agreement conditions. [ref. Ch. 13, 14 and 15]

Make or seek a decision to proceed with drainage work

Drainage work Seek estimates from suppliers of materials and from drainage contractors, select the best offers (not necessarily the cheapest) and arrange contracts.

Supervise the work as it progresses. Stop work in wet conditions and insist on remedial or replacement work as faults are detected.

Make adjustments to design and layout as appropriate when new factors are identified as the work advances.

Observe the site during the next wet season to note the drainage improvement and to identify faults. Improve faults as required.

Glossary

Aquiclude a geological formation which, although capable of absorbing water slowly, will not transmit it rapidly enough to furnish an appreciable supply for a well or spring.

Aquifer a porous geological formation which can store an appreciable amount of groundwater and from which water can be extracted in useful quantities.

Artesian spring a spring which is supplied by water escaping upwards from an aquifer in which groundwater is held under pressure.

Available water that proportion of water in a soil that can be readily absorbed by plant roots (usually recognised as the water held in soil between the conditions of field capacity and wilting point).

Bank slip the detachment of slices of bank-forming material from the sides of ditches or streams by the process of erosion and the influence of gravity.

Base flow the discharge of a surface channel when the soil in its catchment area is in a condition of soil water deficit. The flow is maintained mostly by groundwater escaping from aquifers.

Batter the gradient of a ditch bank.

Bulk density the dry weight of a unit volume of soil in its field condition.

Capillary rise an upwards movement of soil water through fine capillary pores as a result of pressure differential.

Catchment a natural water-collecting area supplying water to a single waterway.

Clay mineral particles in soil whose effective diameters are less than 0.002 mm.

Closed inlet drain drains which receive inflow of water directly or indirectly through the soil profile.

Compaction the result of farming activities similar to natural induration in which the bulk density of soil is increased.

Deep seepage downwards water movement in the soil (and rocks) beyond the range of most plant roots.

Design drainage rate the calculated drainage flow rate based on an appropriate rainfall rate for the area with adjustments made for site characteristics.

Design rainfall rate the selected amount of rainfall in a given time from which the design drainage rate is calculated.

Ditch an open waterway which collects and/or conveys drainage water.

Drainflow the amount of water flowing through a drain in a given time.

Drain layout the pattern of drains installed across the land to deal with a drainage problem in the field.

Drift any type of material deposited by geological processes in one place after removal from another.

Entry resistance the restriction of flow into a field drain imposed by the physical forces which affect water molecules near to narrow entry points in the wall of the pipe.

Evapotranspiration the water loss from soil by the process of surface evaporation plus the water removed by plants transpired to the atmosphere from the leaf surface.

Excess rainfall the difference over a period of time between the gains of water (rainfall, etc.) and losses (evapotranspiration) from the soil.

Field capacity the water content of a soil after excess water has just finished draining by gravity.

Filter material which may be used to coat a drain in order to reduce the possibility of siltation and blockage of the drain.

Flood plain a ribbon of low-lying, flat land beside a river channel which becomes covered with water in times of flood.

Flow line a graphical representation of the flow paths of groundwater towards a drainage channel set in saturated soil horizons.

Glacial till the material deposited by a glacier, consisting of boulders, rocks, gravel and soils of different texture. It is generally unstratified (i.e. well mixed up).

Gley a soil (or horizon) developed under the conditions of poor drainage and usually brownish grey, olive grey or mottled in appearance.

Gradient degree of slope of a surface or a drain and representing the angle between the surface or the drain and the horizontal.

Gravitational water the soil water which is found in the larger pore spaces between the soil particles and drains away under the influence of gravity.

Gravity spring the effect of groundwater seeping from an outcropping aquifer as a result of gravitational groundwater movements.

Groundflow the movement of groundwater in response to gravitational forces.

Groundwater all water contained in soil horizons or rock formations below the level of the groundwater table.

Groundwater table the upper limit of the zone of total saturation in an aquifer.

Head loss (in an aquifer) the difference in height of the groundwater table in different parts of the same aquifer resulting from resistance to groundflow through the aquifer.

Hill drain small, open, surface drainage channels.

Horizon a layer of soil, approximately paralleling the soil surface which has definite visible characteristics produced as a result of soil-forming processes.

Humus the more or less stable fraction of organic matter remaining in soil after the decomposition of residual plant and animal remains.

Hydraulic conductivity the rate at which groundwater flows through a soil in response to a given hydraulic head.

Hydraulic gradient the difference in hydraulic head at two points in a soil divided by the distance between the points measured along the direction of groundflow.

Hydraulic head the difference in elevation of the surface of different parts of a body of water including groundwater in an aquifer. The potential energy may result in flow movements that tend to cancel the hydraulic head.

Hydrologic cycle the continuous circulation of water from the atmosphere through soil to ocean to atmosphere (interrelationships between precipitation, evaporation, ground water supplies, and water in general).

Induration soil material cemented by natural processes into a hard mass which will not soften on wetting.

Infiltration rate the maximum rate at which water will enter the soil under specified conditions (mm/hr or mm/day).

Interflow lateral movement of water through topsoil when infiltration rate exceeds percolation rate.

Iron ochre a reddish-brown gelatinous sludge which is deposited in and around drain-pipes restricting flow.

Land drainage removal of excess, superfluous or gravitational water from soil.

Leaching removal of materials in solution from the soil by the downward passage of water through the profile.

Liquid limit the water content corresponding to the arbitrary limit between the liquid and plastic states of consistency of a soil.

Loam the textural class name for soil having a moderate amount of sand, silt and clay. Intermediate in texture and properties between fine-textured and coarse-textured soils.

Mineral soils a soil consisting largely of mineral materials having less than approximately 20 per cent organic carbon by weight.

Mole drain unlined underground drain channels formed by pulling a bullet-shaped cylinder through the soil.

Open inlet drain drains which receive at least part of their inflow from open channels or surface flow.

Organic soils a soil which contains organic matter in such quantity that this dominates the soil characteristics.

Outlet any channel into which water is diverted to aid land drainage.

Pan horizons or layers in soils that are strongly compacted, indurated, or very high in clay content.

Peak flow maximum likely values of water flow following storm periods.

Ped a soil aggregate formed by natural processes such as a crumb or granular aggregate, etc. (in contrast to a clod, which is formed artificially).

Perched water table the upper surface of a body of ground water in a zone of saturation which is separated by unsaturated material from a body of groundwater below it.

Percolation rate the maximum rate at which water will flow into the subsoil from the topsoil under specific conditions (mm/hr or mm/day).

Permeable fill permeable material (e.g. gravel, stones), usually placed in a drain trench to facilitate water movement into a drain.

Plastic limit the water content corresponding to the arbitrary limit between the plastic and semi-solid status of consistency of a soil.

Poaching damage to the structure of the soil surface caused by smearing and compressing effects of animals hooves. Soils in a wet, plastic condition are more vulnerable to such damage.

Pore space the natural voids in the soil volume between the soil particles which are normally occupied by air and/or water.

Rainfall intensity amount of rain falling in a given time.

Rainfall return frequency the time which elapses between a particular rainfall intensity and its likely recurrence in an area (obtained from rainfall/climate statistics).

Reservoir an open water storage system. Water is stored in a dam or reservoir so that the flow of a stream (or pumping station capacity) is more fully utilised.

Rock permeability the capacity of a rock (or soil) to transmit water. Measured as a velocity (e.g. cm/h).

Rock porosity the fraction of the rock (or soil) volume not occupied by solid particles.

Run-off that portion of incident rainfall which does not enter the soil, but which moves over the soil surface to the surface drainage system.

Sand mineral particles in soil where effective diameters lie in the range 2–0.02 mm.

Silt mineral particles of soil whose effective dimensions lie in the range 0.02–0.002 mm (or a textural class of soil with 80 per cent or more of silt and less than 12 per cent of clay).

Siltation the deposition of water-borne sediment in water channels, lakes, flood plains, etc.; usually as a result of decrease in the flow velocity of water.

Soil capping the orientation and packing of dispersed soil particles which occurs as a result of the effects of rainfall on an exposed surface. The slurried suspension on drying forms a relatively hard crust-like layer at the soil surface.

Soil drainage status the average degree of wetness of a soil.

Soil profile a vertical section of the soil through all its horizons and extending into the parent material.

Soil structure the combination or arrangement of primary soil particles into aggregates (or peds) which, together with the associated pore spaces, constitutes the physical framework of the soil.

Soil texture the relative proportions of the various soil separates (i.e. sand, silt and clay) in a soil.

Soil water deficit the quantity of water required (mm) to restore the moisture content of a soil to field capacity.

Soil water tension the equivalent negative pressure or suction in the soil moisture. Expressed in pressure unit it is the force/area that must be exerted to remove water from soil.

Spring sites where groundwater emerges at the soil surface.

Static head distance water must be elevated between the surface of water in a pump and the outlet channel.

Stream load all rock fragments and dissolved substances from rocks moved along in the stream water.

Subsoiling a tillage operation primarily concerned with altering the soil structure below normal depth of ploughing. The objective is to open up compacted soil layers by dragging a tined implement through the subsoil.

Superfluous water excess or gravitational water which lingers in the soil to the detriment of plant growth and agricultural operations.

Surcharging where water inside a pipe is under pressure because the quantity exceeds the pipes capacity to carry it.

Surface flow the movement of free water over the surface of (sloping) land.

Surface water problem a soil drainage problem caused by layers or horizons in soil which restrict the downward movement of soil water (giving rise to a 'perched water table').

Topography shape and physical features of the land surface that make up the landscape of an area.

Underdrainage removal of excess water from the soil by a subsurface system of drainage.

Water neutral surface a land surface which causes surface flow along parallel lines.

Water-gathering surface a land surface in which surface run-off of water tends to converge towards a definite site.

Watershed the line of separation between adjacent water catchment areas.

Water-spreading surface a land surface in which surface water run-off tends to flow along diverging lines.

Weathering the processes involving chemical, physical and biological activity which causes the conversion of rocks into soils.

Wilting point the water content at which plants growing in the soil wilt due to lack of available water.

Index

Page numbers are printed in italic where an entry is defined for the first time in the text

actual evapotranspiration *14*
air mass *89*
alluvial fan *31*, 122
alluvium *105*, 106, 107, 231, 232
aquiclude *109*, 114, 116, 117, 218, 225, 242
aquifer *109*, 116, 117, 118, 119, 126, 189, 226, 227, 242
artesian spring *118*, *120*, 121, 124, 125, 126, 189, 227, 228, 229, 231, 242

bank slip *136*, 242
bank stabilisation *138*, 209
base flow *101*, 242
base level *103*, 105
basin mire *67*
blanket mire *66*
braided stream *104*, 105
brown forest soil *55*, 91
bulk density *52*, 242

capillary pore *16*
capillary rise *17*, 242
carrier ditch 130, 194, 211
carrier drain *147*, 164, 212
catchment *99*, 101, 200, 201, 242
channel rugosity *100*
chemical weathering *29*, 35
clay *36*, 42, 125, 176, 177, 179, 180, 181, 183, 218, 242
clay pan *51*

closed-inlet drain *157*, 242
collector ditch *130*, 211
collector drain *146*, 156, 178, 179, 188, 213
composite layout *211*, 216, 217
concentration loss *153*
confined aquifer *115*, 121, 124, 227, 228, 229, 230
confined groundwater table *115*, 137
convectional rainfall *90*
critical depth 177, *183*, 184
critical season *93*, 95, 148, 149
culvert *139*
cyclonic rainfall *90*

deep seepage *5*, 148, 243
degree of humication *73*
delta *82*, 122
design drainage rate *147*, 148, 151, 152, 153, 156, 240, 243
design rainfall rate *95*, 96, 132, 148, 153, 200, 240, 243
dipping aquifer *116*, 117, 119
discharge *98*, 131, 132, 134, 135, 153, 156, 158
ditch *128*, 243
  batter *136*, 242
  cleaning *140*
  depth 131, *134*, 137
  maintenance *143*
  piping and filling *142*
  regrading *140*

re-planning *143*
re-siting *141*
straightening *141*
drain *145*
  blow-out *155*
  depth *169*, 217, 224
  layout 210, *211*, 243
  maintenance 173, 174
  spacing *217*, 221, 222, 223, 225
drainage coefficient *147*
drainflow *147*, 148, 149, 152, 154, 240, 243
drift *31*, 122, 123, 125, 239, 243
dyke *28*, 120

easy-pathway spring *119*, 118, 123, 226
entry resistance *153*, 154, 243
equivalent depth *221*
evapotranspiration *4*, 7, 148, 200, 243
excess rainfall 7, *94*, 98, 243
excess soil water 63, 92, 147, 153

fen layout *214*
field capacity *15*, 243
field drain *145*, 163
filter *187*, 235, 236, 243
flash flood *102*
flood plain *104*, 105, 106, 122, *231*, 243
fluvioglacial drift *33*, 120

geomorphology 79
glacial till 32, 119, 123, 124, 243
gley 56, 243
    horizon 56
    soil 91
gradient 85, 133, 135, 137, 149,
        154, 166, 173, 178, 192, 216,
        231, 239, 243
gravitational water 15, 16, 109, 244
gravity spring 118, 125, 226, 231,
    244
grid layout 214, 217
groundflow 6, 108, 112, 119, 126,
        132, 148, 150, 152, 160, 201,
        218, 224
groundwater 3, 108, 118, 119, 123,
        126, 227, 239, 244
    flow line 113, 122, 243
    gley 62
    peizometric surface 116, 117,
        126
    phreatic surface 115
    table 5, 7, 110, 111, 112, 113,
        160, 244
    table control ditch 129, 212
grus 36, 124

head loss 111, 244
herringbone layout 215, 217
high groundwater table 11, 88,
        126, 218, 219, 223, 231
hill drain 191, 192, 194, 244
humose soil 65, 71
humus 54, 244
hydraulic conductivity 52, 60, 219,
        222, 223, 239, 244
hydraulic friction 155, 156
hydraulic gradient 100, 111, 116,
        154, 244
hydraulic head 100, 244
hydrologic cycle 3, 105, 244

individual layout 213
indurated horizon 51

indurated profile 51, 185, 186, 218,
    224
infiltration rate 5, 114, 149, 244
interceptor ditch 130, 194, 212
interceptor drain 146, 156, 160,
        189, 213, 226, 227, 228
interceptor layout 214
interflow 6, 150, 244
intermittent spring 118
iron ochre 53, 233, 234, 236, 245
iron pan 51
irregular layout 213
irrigation 8, 63

kaolinite clay 37, 179, 180

land drain 145, 146
land drainage 8, 63, 156, 194, 245
    ditch 129
lateral drain 146, 213
leader drain 146, 164
loam 41, 42, 218, 245
local base level 103, 198, 231

main drain 146
marginal bluff 104
mass wasting 80
mature landscape 83, 86
mature sediments 31
meltwater channel 105
micaceous clay 37, 181
mineral soil 39, 245
minerotrophic mire 67
minor drain 146
mole drain 175, 177, 179, 180, 245
mole drainage 176, 187, 188, 195
mor humus 66
mull humus 66

natural layout 213
natural pan 51, 185

old landscape 83, 87
ombrotrophic mire 68
open-inlet drain 158, 245
organic soil 64, 245
orographic rainfall 90
outlet 98, 160, 168, 240, 245
overtopped-barrier spring 118,
        120, 121, 124, 226

pan 51, 52, 184, 185, 224, 245
peak flow 101, 240, 245
peat 57, 65, 72, 181, 232, 233
    humification 73
    oxidation 76
    shrinkage 75, 76, 232
peaty gley 57
peaty podzol 57
peaty soil 65, 72
ped 45, 46, 47, 48, 49, 245
perched water table 11, 175, 218,
        224, 225, 245
percolation rate 5, 149, 150, 245
permanent spring 118
permeable fill 176, 180, 187, 188,
        189, 195, 224, 225, 231, 233,
        245
permeable rock 27, 109, 246
physical weathering 29, 34
pipe size 153, 154, 156, 157, 158,
        159, 160
pipe type 163, 164
plough pan 23, 52, 184
podzol 55, 91
potential evapotranspiration 14
pump
    drainage 198, 240
    head loss 205
    reservoir 201, 209, 246
    static head 201, 247
    sump 201, 228
    total head 205

quaking mire 70

rainfall *2*, *89*, 90, 148, 200
  amount 90, 91, *93*, 200
  duration 94, 95
  intensity 6, *94*, 114, 149, 246
  return frequency *93*, 94, 96, 246
rain shadow *90*
raised mire *68*
regular layout *214*
return flow 7, 201, 231
rillflow *81*
rock
  bedding plain *27*
  fault *27*, 119, 120
  fold *27*
  formation *26*, 27, 28, 108, 109, 116, 239
  joint *27*
  permeability *27*, 109, 246
  porosity *108*, 246
  structure 26, *27*
  waste *30*
run off *99*, 101, 131, 148, 150, 151, 153, 246

sand *36*, 42, 218, 246
sedentary rock waste *31*
semi-confined aquifer *116*, 122, 227
sheet flow *81*
sill *28*, 119
silt *36*, 42, 218, 235, 246
siltation *135*, 152, 153, 187, 246
singular layout *211*, 216
smectite clay *37*, 179
soil 4, 12, *39*, 64
  aggregate *45*
  A horizon *40*
  air *12*, *15*
  association *57*
  B horizon *40*
  capping 45, *52*, 184, 246
  C horizon *40*
  classification 54, 64, 239
  compaction *52*, 182, 242

drainage status *18*, 24, 246
  horizon 5, 40, 71, 239, 244
  induration *51*, 182, 185, 244
  leaching *51*, 245
  liquid limit *46*, 245
  macropore *16*
  mesopore *16*, 149
  micropore *16*
  natural drainage *18*
  parent material 5, 40
  plastic limit *46*, 246
  poaching *23*, 52, 246
  pore space *15*, 50, 246
  profile 5, 39, 58, 59, 125, 218, 239, 246
  series *57*
  structure 45, 46, 48, 50, 75, 115, 246
  texture *41*, 45, 50, 74, 218, 247
  treatment *63*, 175, 224, 225
  water *3*, 12, *15*, 16, 17, 98
    availability *17*, 44, 242
    content *15*
    deficit *15*, 17, 91, 179, 247
    tension *16*, 247
solifluxion *80*
  drift *34*
solum *40*
spoil *138*, 172
spring 6, 118, 194, 218, 247
springline *118*, 194
spring mire *70*
spring mound *121*, 124
storage ditch *130*
stream abandoned meander *104*
stream bed load *102*
stream dissolved load *102*
stream flow 4, 98, 99, 100, 101
stream load *102*, 247
stream meander *104*
stream suspended load 102
sub-leader drain *146*
sub-main drain *146*
subsoil *5*
subsoiling *176*, *181*, 182, 184, 185,

186, 187, 247
superfluous water *16*, 247
surcharging *154*, 155, 247
surface
  aquifer 109, 110, 111, 115, 160, 218, 219, 223, 224, 239
  flow 6, 131, 150, 218, 247
  inflow *11*, 150, 195, 200, 203
  water *98*
    gley *62*
    problem *11*, 247

time of concentration *101*, 102
topography 7, *79*, 149, 212, 247
topsoil *5*, 239
transpiration *3*, 13
transportation 81, 102
trenchless drainage *163*, 188, 189

unconfined aquifer *115*
underdrainage 145, 186, 195, 247

water-gathering surface *85*, 150, 151, 247
water neutral surface *85*, 150, 151, 247
watershed 99, 247
water-spreading surface *85*, 150, 151, 247
weathering 29, 36, 82, 247
wilting point *15*, 17, 247

young landscape *83*, 87

zone of intermittent saturation *110*, 111
zone of permanent non-saturation *110*
zone of permanent saturation *110*, 111